# ARCHITECTURAL DRAFTING
# ASSIGNMENTS USING
## AutoCAD

# ARCHITECTURAL DRAFTING
# ASSIGNMENTS USING
## AutoCAD

**Myron R. Brower**

## autodesk Press

THOMSON

DELMAR LEARNING

Australia • Canada • Mexico • Singapore • Spain • United Kingdom • United States

**THOMSON**

**DELMAR LEARNING**

**autodesk** Press

## Architectural Drafting Assignments Using AutoCAD
Myron R. Brower

**Autodesk Press Staff**

Vice President,
Technology and Trades SBU:
Alar Elken

Editorial Director:
Sandy Clark

Senior Acquisitions Editor:
James DeVoe

Senior Development Editor:
John Fisher

Marketing Director:
Dave Garza

Channel Manager:
Dennis Williams

Marketing Coordinator:
Stacey Wiktorek

Production Director:
Mary Ellen Black

Production Manager:
Andrew Crouth

Production Editor:
Stacy Masucci

Editorial Assistant:
Tom Best

For more information contact
Delmar Learning
Executive Woods
5 Maxwell Drive, PO Box 8007,
Clifton Park, NY 12065-8007
Or find us on the World Wide Web at
www.delmarlearning.com

For permission to use material from the text or product, contact us by
Tel.   (800) 730-2214
Fax   (800) 730-2215
www.thomsonrights.com

Library of Congress
Cataloging-in-Publication Data

Brower, Myron R.
  Architectural drafting assignments using AutoCAD / Myron R. Brower.
    p. cm.
  ISBN 1-4018-9031-8
  1. Architectural drawing–Computer-aided design. 2. AutoCAD. I. Title.
  NA2728.B756 2005
  720'.28'40285536–dc22
                            2004030465

### NOTICE TO THE READER

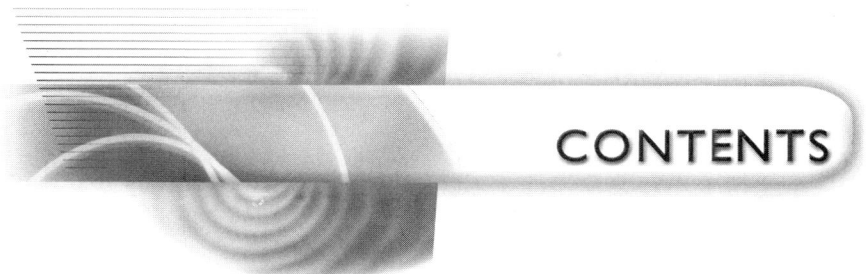

# CONTENTS

## GROUP 1: GEORGIAN HOUSE

### ASSIGNMENTS:

## GROUP 2: BROWNSTONE

### ASSIGNMENTS:

## SUPPLEMENTS

<br />
<br />

## PREFACE

### INTRODUCTION

As the program director for Architectural Technology/CAD at Scottsdale Community College and, perhaps more importantly, as a classroom instructor, I am convinced that a student's first semester of Computer-Aided Drafting is in many ways the most challenging. For most students, everything—yes EVERYTHING—is completely new. The work in this text presents a series of interconnected assignments that *logically* and *cleanly* build basic two-dimensional (2D) CAD skills within architectural conventions and formats typically employed by architects, interior designers, building contractors, and related consulting engineers. Despite this particular focus, I believe that this work will provide a solid introduction to the 2D CAD skills required by most instructional programs or professions.

These assignments are time proven, honed through years of classroom application, student feedback, and instructors' refinements. The work "cuts to the chase," helping students master basic 2D skills without wrestling with the comprehensive, intimidating—and often irrelevant—exercises typical of most general CAD tutorials.

Our SCC faculty continue to be amazed at the distance our first semester students travel over the course of these assignments. The students themselves are typically awestruck with their success. As each assignment unfolds, as students' intuition and confidence grows, and as students begin to develop working relationships with one another, faculty move from their early role of leader/demonstrator to that of a wandering trouble shooter. The learning experience becomes self-sustaining; students learn "how to learn" and how to solve problems on their own.

These assignments may be used in a variety of ways: as stand-alone work designed for use with classroom lectures and instructor demonstrations, in combination with instructor presentations and so-called "Quick Reference Guides" such as the excellent work of Ralph Grabowski's (as is our current practice at SCC), or in supplement to other textbooks and/or tutorials.

The assignment structure is geared for AutoCAD 2004 (AutoDESK™), but its generic base allows it to work with other releases, requiring only slight adjustments in instruction and navigation.

### BREATHE, MY CAD GRASSHOPPERS...
### AND READ THIS WHILE DOING SO

Nervous? Take a deep breath. Breathing on a regular basis has proven to be a good thing, and while we're at it, it's also generally considered a good idea to *exhale* after taking those nice deep breaths. Next, rest easy. Most beginning students are a bit uneasy anticipating their first encounter with Computer-Aided Drafting (CAD). What can I, as a teacher and veteran of many such campaigns, say to you as a beginning student? Several encouraging words come to mind: challenging, amazing, and rewarding; the obstacles: frustration, confusion, and intimidation; the tools to overcome those obstacles: persistence, objectivity, and patience ... oh yeah—and oxygen—that breathing thing, again.

For most new students, even those with "other" computer expertise, the operations and processes you are about to learn are totally new. You may likely have no previous learning base on which to build. So the learning curve will start out a bit flat. The trick, then, will be to cut yourself some slack. Be patient with your own progress; seek help from and offer help to your fellow students; *ask your instructor questions;* and most of all, PUSH THE BUTTONS! No matter how many times you *hear* how an operation is performed, how many times you *watch* a demonstration, how many nights you spend *reading* textbooks, the key to learning CAD lies within yourself and your resolve to spend the time at your work station, pulling down menus, selecting icons, keyboarding, and "left and right clicking" your way toward a completed drawing.

Don't be afraid to make mistakes; that's why the computer gods invented the UNDO and ERASE commands. I believe that often the best and most lasting learning comes through correcting a mistake. Be philosophic and objective—you're learning here—things will happen for reasons you will be unable to explain. There's even an outside chance that you may send a drawing or two to AutoCAD heaven, never to be seen again. Learn from such experiences and stay objective. In the big picture, losing a drawing of a Shutter is not like you've dumped thirty years' worth of research, calculations, and drawings for the Space Shuttle. In fact, the lessons you learn from your beginning errors may prevent you from encountering real catastrophes somewhere down the road.

Don't be hard on yourself. With persistence, patience, and an objective attitude, that early, flat learning curve will soon make a dramatic and rewarding spike in the "up" direction. Also, every once in a while, sit back from your screen and think. Remind yourself how much you've learned to that particular point, consider how many commands, once so new and difficult, are now second nature. That is to say, every now and then—while you're breathing—take some time to pat yourself on the back. Multitask!

Stick with this stuff. In a short time you will stand in awe of your work and the enormous distance you have traveled over the course of these assignments.

## HOW TO USE THIS TEXT

**Learn, don't copy.** The following assignments are carefully designed and logically planned to build on each other while the student builds his or her 2D AutoCAD skills. If one underlying premise threads its way through the work, it's the presumption that your skill development and eventual **mastery will come through repetition.**

In that light, the instructions for each assignment do not necessarily lead you through the exercises in the most efficient manner. Rather they are purposely designed to reinforce—through repetition—specific command(s) new to the particular assignment or introduced in previous assignments. It then becomes paramount for true success with this work that you **follow instructions,** using the commands as directed, especially so for the first groups of assignments.

In a very short time, most students have developed enough expertise to "draw the picture" by "just looking at the picture." This is a slippery slope if *learning* is your goal (which, by the way, IT IS!). Make sure that you don't outsmart yourself. Read and follow directions, use the new commands designated in **bold text** for each progressive assignment. *Ask questions of your instructor,* of fellow students, and/or refer to supplemental texts if you are asked to perform a command that is unfamiliar. Don't shortcut your way around the obstacle.

**Make things easier.** The cover page for each assignment lists the commands and operations *du jour.* Note that commands and/or operations new to a particular assignment are printed in **bold text.** Again make sure that you read supporting information and use the new commands. Don't skirt the lesson's objective(s).

You may find it to your advantage to **remove the example drawings** for each unit to avoid "flipping" back and forth between the written instructions and the illustrations. We've perforated the text for that very reason.

**Assignment groups:** Assignments form two major groups: the **Georgian House** and the **Furness Brownstone.** Thirteen separate assignments make up the **Georgian House.** Over this group you will be seeing commands for the first time and be given the opportunity, again through repetitive use, to begin your mastery of the most typical 2D operations. All thirteen assignments culminate in a final plot of a dimensioned and noted Floor Plan and Elevation incorporating all the previous assignments. For example, your first assignment is to draw a Door. That Door will eventually be inserted into your final drawing of the exterior elevation, as will be the case for your second assignment, the Window.

The second group of assignments, the **Furness Brownstone,** involves only a few new operations and commands. While the drawings themselves are more complex, this group allows you to practice commands first introduced in the Georgian group. Speed and confidence build with this group, and it's here that you start to develop your own creative tacks, navigating through what are now the not-so-mysterious waters of AutoCAD.

And...when completed successfully, the final plots are real knockouts!

Myron R. Brower

GROUP

1

GEORGIAN
HOUSE

# Panel Door

## COMPETENCIES/LEARNING OUTCOMES

Upon successful completion of the **Panel Door** the student will have used and begun to master the following settings, operations, and drawing commands:

**Settings:** Units, Drawing Limits (Area/Paper Size), Snap, Grid, Ortho

**Operations:** Open/Exit AutoCAD; Access Menus; Enter Commands; Startup Wizard; Open; Save As (name a file) and Save drawing files

**Commands:**

Line

Offset

Fillet (0)

Erase: Select, Window, Crossing Window

Undo

Zoom Window, Previous, All, Realtime

Pan

Save

Save As

Plot in Paper Space

## PROCEDURE

Working alone and with classmates, incorporate information from lectures, other text(s), demonstrations, and/or corresponding information to complete this drawing. In this, your first drawing, be patient and have fun getting to know the computer. While there are many ways to access individual commands, find the one that seems the easiest for now—don't worry about remembering everything all at once. Your expertise will come with hands-on practice ... not from memorizing information.

**RELAX, EXPLORE, ASK QUESTIONS,**

**WORK WITH YOUR FELLOW CLASSMATES.**

**Setup:** (Verify with instructor Startup mode = 1)

From the **Create New Drawing** dialog box—this will appear when you first open AutoCAD ...

or ... select **File/New** from the Pulldown menu at the top left of your screen:

Pick    **Use a Wizard:** the icon "with the little stars"

Pick    **Quick Setup** then **OK**... or ... double click **Quick Setup**

Set     **Units:** Pick **Architectural Units**

Pick    **Next**

Set     **Area:** Width = 8'6, Length = 11'

---

### ENTERING DIMENSIONS IN AUTOCAD; PAPER SIZE (AREA)

Note that you do not have to type the inches symbol ( " ) or the typical "dash" between feet and inches. If you just type "numbers," AutoCAD will assume you mean "inches." If you want to enter distances in "feet," then you *must* enter the foot symbol ( ' ).

As part of a typical setup, you will decide on the width (horizontal or "*x*" direction) and length (vertical or "*y*" direction) of the "paper" on which you will draw. The computer typically thinks it is drawing things full size, so you typically want to create a sheet of paper "bigger" than the object you wish to draw. The sizes given tie to the paper size and scale of the final plot (see Supplement 2: Drawing Area/Limits).

---

Pick    **Finish:** You're almost ready to draw! Scary, isn't it?

---

### UNITS AND DRAWING LIMITS—AN ALTERNATE SETUP

FYI ... Another way to set Units and Drawing Area (Drawing Limits) is to select **Start from Scratch** at the Dialog Box and go to the **Format** Pulldown menu at the top of your screen. *Always set Units first*, then set Drawing Limits using *x* and *y* coordinates. [(0,0) typically for the lower left corner, the width (*x*) and length (*y*) for the upper right corner.] You can readjust these settings at any time through the **Format** Pulldown menu options.

---

We may now assign this drawing a name using the **SAVE AS** command. We can then update the drawing by simply using the **SAVE** command. *Verify the location of your "Save As" with your instructor—this is important—some drive locations are automatically cleared—ERASED—at the end of your work session!*

So ... from the **File** Pulldown at the top left of your screen:

Pick    **Save As:** The **Save As** dialog box will appear.

**Save In** box: Scroll to/select the drive location designated by your instructor.

**File Name** box: Delete the default drawing name (typically "Drawing 1");  enter a "new" name for the Door—your three initials**DR**—ex: **mrbDR**

Pick    **Save**

You have now renamed the default drawing, SAVING it AS something "different"—**Saving** it **As** \*\*\*DR. Now that you have given the drawing its proper name, you can merely do SAVE commands (not SAVE AS) to update this drawing file, selecting **Save** in the **File** pulldown or the **Save** icon.

**Other Settings:**

    **Snap** ON (4"). **Grid** ON (4") **(Tools Pulldown/Drafting Settings/Snap and Grid)**

    **Ortho** ON to drag and type Lines.

    Use toggles to turn these settings on and off as required/desired while drawing.

    Set **Fillet Radius to 0"**—always verify Fillet Radius with each new drawing.

       **Remember to SAVE periodically while working and SAVE before you exit.**

## OK, LETS DRAW

Refer to the drawing of the Panel Door. You may wish to remove the drawing example from your text for easier coordination with the written instructions.

1. **Line** command: Draw door outline, 3'-0" × 6'-8".

   With Ortho ON—pick start point, drag in the direction of line, type *and enter* distance to complete each segment of basic outline in one operation.

   **Zoom Window** around door for "bigger view."

   You will need to Zoom "in and out" many times during this process—give your eyes a break—*Zoom Window, Previous, All,* and *Realtime.*

   Use the *Pan* command and scroll bars as well … get to know these tools.

2. **Line** command: Draw 4"-wide jambs (side frames) to the outside of the door.

3. **Line** command: Draw outlines of panels, using 4" Grid points as guides.

   Your Snap and Grid will let you do this accurately.

4. **Offset** ³/₄" panel bevels. Snap OFF—to ease the operation.

   **Verify Fillet Radius = 0"—always verify Fillet Radius.**

   **Fillet** corners, **Zoom** in (Window) and out (Previous) as required.

   You must turn **Snap OFF** to select the lines for Fillets—the offset lines are not on your 4" Snap.

5. **Offset** ³/₈" panel bevels. **Zoom/Pan** as required.

   **Fillet** corners, **Zoom/Pan** as required.

6. Reset Snap to 1", Grid to 2" (Tools/Drafting Settings):

   **Line** command: Snap **ON**—Draw 46 × 6 header and 48 × 2 trim at the top of the door.

7. **Line** command: Draw fluting and base/cap details at jambs.

   Offset 2" for fluting starts/stops—erase offsets when done.

   Draw one vertical flute line using Snaps, then **Offset** the remaining two (each side).

8. Pediment, Crest, and Doorknob are optional …

   Experiment with Circles and Snap settings, Ortho ON and OFF.

**SAVE this drawing before you exit.**

## MORE ON SAVE AND SAVE AS

To save this drawing under its "current name" and to the "current drive" (as indicated by your instructor), just "do a **Save**"—not a "Save As." To **Save** a drawing with **no changes** to its name or drive (location)—pick the **"diskette" icon** at the top left of your screen … or … from the **File** pulldown, pick **Save.**

### SAVE AS AND SAVE

One more time … use **Save As** when you want to save the drawing "As something different," i.e., "As" a drawing with a different name, "As" a drawing on a different drive, or "As" a drawing with a different name AND on a different drive.

Use **Save** to "update" a drawing—with its current name and in its current location.

## PLOTTING THE PANEL DOOR

What a wonderful feeling to see the plot of your first drawing! But first, a few things about plotting that sound much more complicated than they actually are. You have been drawing in an environment AutoCAD calls **Model Space** on "giant" pieces of paper. In Model Space, AutoCAD believes all entities to be "full size"—that is drawn to a scale of 1:1. For the exercises in this text, you will need to plot on "normal" paper sizes and to an architectural scale. Though plotting to scale from Model Space is possible, most offices plot drawings in an environment known as **Paper Space.** In Paper Space, we "see" the actual size of the final plot, i.e., the actual size of our paper, and the *scaled* view of our finished drawing.

You can travel between **Model Space** and **Paper Space** by selecting the **Model** or **Layout Tab(s)** directly at the bottom of your drawing area, or by piloting the Starship Enterprise at warp 19 through a worm hole to an alternate universe ... just beware of Romulan Birds of Prey decloaking in the Neutral Zone. Most believe the former choice to be the more practical.

The primary operations for plotting in **Paper Space** revolve around **Page Setup.** Via that dialog box, you will select a plotter, assign lineweights by color (although some prefer an alternate way of assigning lineweights), and select paper size and orientation.

Your instructor will walk you through the process, and you will probably need some repeated help and practice to master the ins and outs of **Page Setup**, creating and manipulating Viewports, and plotting in Paper Space. Paper Space may appear daunting at first blush, but as is the case with many operations in AutoCAD, the "explanation" is far more complex than the "application." Be extra patient here, and don't be shy about asking questions as you plot this and future assignments. Obviously, learning to plot your drawings is very important.

*Supplement 6: Quick Guide to Plotting in Paper Space* is included this text's Supplements. After a few walk-throughs with your instructor, this guide will begin to make sense. We'll start out easy—not worrying about lineweights, only producing a "check plot."

Refer to the Quick Guide (as needed) as your instructor takes you through the process or as you follow the following "simplified steps" for your first plot.

**Select a Layout Tab** to enter Paper Space:

| | |
|---|---|
| **Page Setup:** | Right click on Layout Tab for Page Setup (typically appears by default). |

Try these settings for the Door:

| | |
|---|---|
| **Plot Device Tab:** | (at top of dialog box) |
| **Plotter Configuration:** | per your instructor |
| **Plot Style Table:** | monochrome—for this plot, don't worry about editing (Edit) lineweights—we'll let the lineweights "default." |
| **Layout Settings Tab:** | (at top of dialog box) |
| **Paper Size:** | letter (8$\frac{1}{2}$ × 11) |
| **Orientation:** | portrait |

Other settings are probably OK ... your instructor will verify.

**OK** Page Setup and you'll see your designated piece of paper.

| | |
|---|---|
| **Load the Viewport Toolbar:** | Right click on any Toolbar and select **Viewport**. Create a single Viewport nearly the size of your paper boundaries. |

(A Viewport is typically created automatically—you may wish to adjust its size and location using Grips, or Erase it and create a new single Viewport.)

**Activate the Viewport:** "Activate" the Viewport by double left clicking "inside the Viewport" or by toggling from Paper Space back into Model Space in the Drawing Aids (Snap, Ortho ... group) at the bottom of your screen. Although the screen looks the same, this moves the image to Model Space.

**Scale** the Viewport to 1" = 1'-0" using the Viewport Toolbar.

Center the Door in the Viewport (**Pan**— "Mr. Hand").

**Deactivate the Viewport** (return to Paper Space): Double left click "outside the Viewport," or toggle at the bottom of your screen.

**Place your name in Paper Space**: A preview of the **Text** command:  Type **DT** (for Dynamic Text) and **Enter.** Follow prompts. Text height ⅛".

**Save** your finished setup.

**Right click** on the Layout Tab and select **Plot ... OK.**

**OK!**

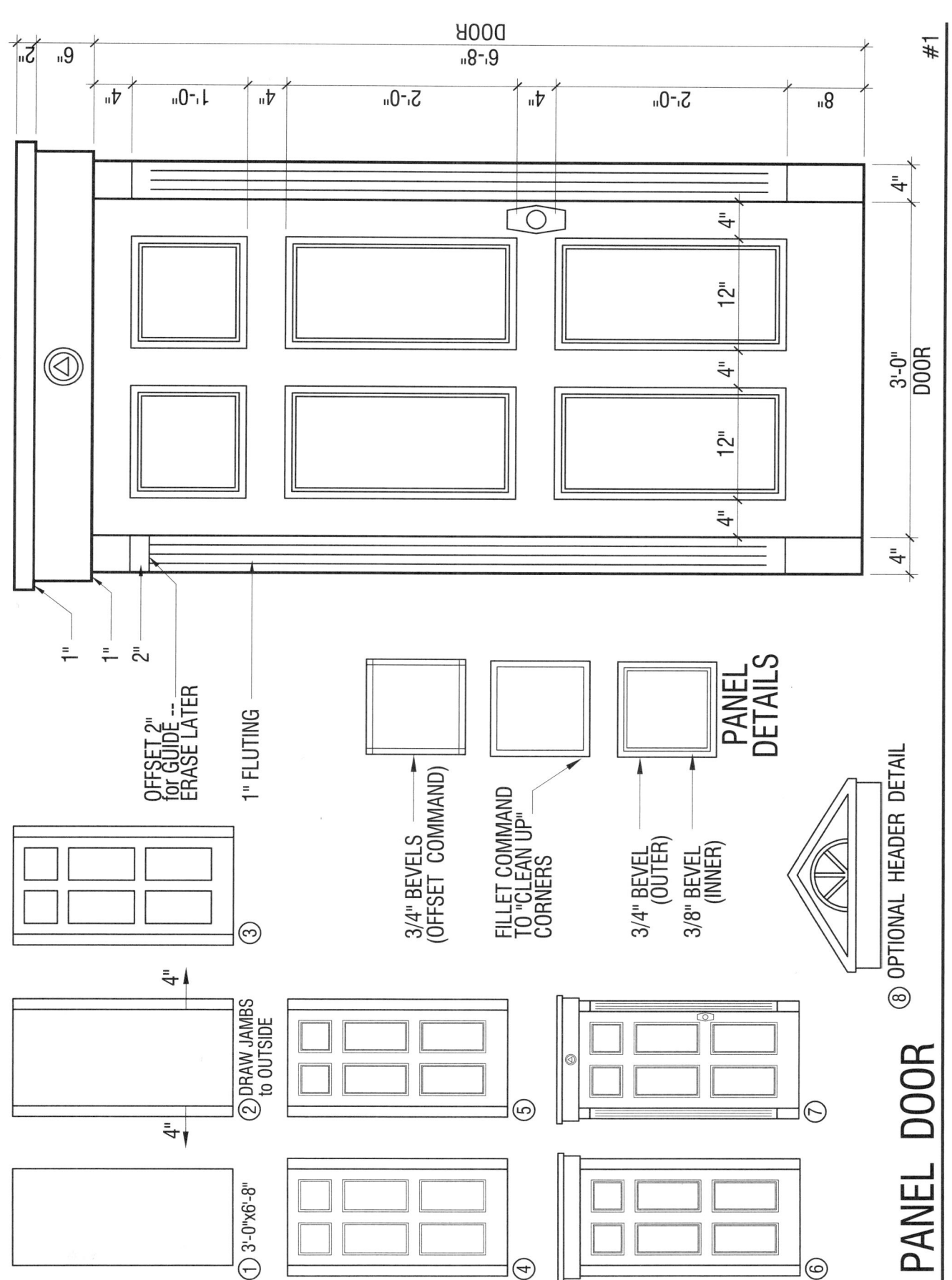

DOOR

6'-8"

2"
6"
4"
1'-0"
4"
2'-0"
4"
2'-0"
8"

1"
1"
2"

OFFSET 2"
for GUIDE --
ERASE LATER

1" FLUTING

4"
12"
4"
12"
4"
4"
3'-0"
DOOR
4"

3/4" BEVELS
(OFFSET COMMAND)

FILLET COMMAND
TO "CLEAN UP"
CORNERS

3/4" BEVEL
(OUTER)

3/8" BEVEL
(INNER)

PANEL
DETAILS

③

④

⑤

⑥

⑦

⑧ OPTIONAL HEADER DETAIL

4"

4"

② DRAW JAMBS
to OUTSIDE

① 3'-0"x6'-8"

# PANEL DOOR

#1

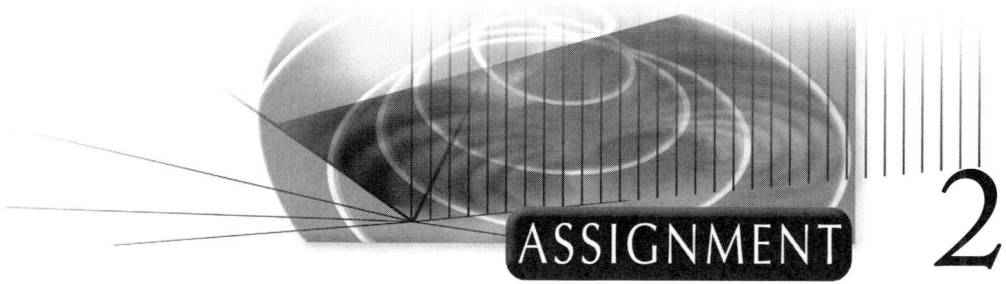

<br />
<br />

# ASSIGNMENT 2

## Window

<br />

## COMPETENCIES/LEARNING OUTCOMES

Upon successful completion of the **Window,** the student will have used and begun to master the following settings, operations, and drawing commands.

**Settings:**   Units, Drawing Limits, Snap, Grid, Polar Tracking, Ortho, OSnap

**Operations:** Open/Exit AutoCAD; Access Menus; Enter Commands; Startup Wizard; Open; Save As and Save drawing files; **Creating/Saving backup files**

**Commands:**

    Line

    Offset

    Fillet (0)

    Erase: Select, Window, Crossing Window

    Undo

    Zoom Window, Previous, All, Realtime

    Pan

    Save

    Save As

    Plot in Paper Space

(new)   **Trim**

    **Object Snap (OSnap) (midpoints)**

## PROCEDURE

Working alone and with classmates, incorporate information from lectures, demonstrations, and corresponding information in your text to complete this drawing. In this, your second drawing, you will continue to reinforce lessons/commands encountered in your previous drawing, the Panel Door. Look for similarities in applying the various commands as well as between the creation of this drawing and your earlier assignment. Again, be patient, pat yourself on the back for completing the first assignment and jump into this one. Remember that expertise comes with practice.

**RELAX, EXPLORE, ASK QUESTIONS,**
**WORK WITH YOUR FELLOW CLASSMATES.**

<br />

<br />

<br />

<br />

9

> ## VERIFY YOUR SAVE AS/SAVE DRIVE LOCATIONS!
>
> Take care that when you **Save As** and/or **Save** your drawing that you save to the location designated by your instructor. In this exercise you will also be directed to save a backup copy of this drawing on the (A:) Drive (Floppy Drive). IN SOME INSTANCES, DEFAULT "GUARDS" MAY BE SET AT YOUR WORK STATIONS THAT AUTOMATICALLY ERASE WORK WHEN YOU SHUT DOWN! IN THOSE CASES, *YOU WILL LOSE YOUR DRAWING* IF YOU INADVERTENTLY SAVE TO A PROTECTED DRIVE LOCATION. **Ask your instructor if you have questions.**

## SET UP A *NEW* DRAWING

If you are in the middle of a work session and finish a drawing, **Save** it one last time, then, from **File** Pulldown, select **Close**. To then begin a new drawing, from the **File** Pulldown, select **New** (or select the appropriate icon). The **Create New Drawing** dialog box will appear and off you go!

Pick     **Use a Wizard:** the icon "with the little stars"

Pick     **Quick Setup** then **OK**

Set      **Units:** Pick **Architectural** Units

Pick     **Next**

Set      **Area:** Width = 11', Length = 8'6 (a "landscape" setup—long side down)

Pick     **Finish**

**Zoom ALL** to ensure that you have captured your entire screen.

Pick **Save As** (from **File** Pulldown) to access the **Save As** dialog box

**Save In:** Locate your designated drive—verify with your instructor.

**File Name:** your three initialsWIN—ex: mrbWIN or mlkWIN

(Remember to **SAVE** periodically while working and **SAVE** **before you Exit** AutoCAD)

**Other Settings:**

**Tools/Drafting Settings: Snap and Grid** tab: Snap 2"; Grid 4"

                 **Object Snap** tab: verify **OSnap** Settings:

                              Endpoint, Midpoint, Center, Intersection, Perpendicular ON

                              Other OSnap options OFF.

                              Always verify these settings with each new drawing!

Verify **Fillet Radius = 0"** Always verify Fillet Radius with each new drawing!

**Toggles: Snap** ON; **Grid** ON; **Ortho** ON; **OSnap** ON

       You may change these settings and/or turn them on and off as required while drawing.

                 **SAVE . . . and remember to SAVE periodically while you work.**

## OBJECT SNAP . . . A NEW DRAWING AID

Object Snaps (OSnaps) rule! When OSnaps are operating, AutoCAD will automatically lock onto designated points of existing entities—an Endpoint or Midpoint of a line, the Center of a Circle, an Intersection of two entities, a point on an entity perpendicular to another entity—among many other options. You may access OSnaps several ways—by typing the desired OSnap during a command, by loading and docking the OSnap toolbar, or more typically by setting "Running Object Snaps" using the Drafting Settings Dialog box (which you just did as part of your Setup). We use the term "Running" OSnaps when the OSnap option is defaulting—or "running"—all the time. You can activate (or deactivate) the OSnap option by using the OSnap toggle at the bottom of the screen.

Your instructor will demonstrate OSnaps and their use. Watch and be amazed!

Four things to remember when using Running OSnaps:

- **Don't confuse OSnap with the "plain old SNAP" setting.** As you progress through your assignments, you will rely less and less on "SNAP" and more and more on "OSnap" for speed, flexibility, and accuracy.

- The OSnap command *ensures accuracy* when drawing—don't ever try to "eyeball" corners and the like with AutoCAD as we're no match for AutoCAD's built-in accuracy. Eyeballing in AutoCAD is the most slippery of slopes—try it and you'll find yourself in a real mess when it comes time to dimension your drawing.

- You only need to be "close" to the point you're after. When you've maneuvered close to the target, a symbol will appear at the desired location. **Learn to recognize the symbol for each OSnap—** square, triangle, x, etc. Sometimes a Perpendicular point is very close to a Midpoint; you need to make sure that you OSnap to the target you desire.

- **The OSnap is your friend—***almost* **always.** While you will typically draw with the OSnap toggle ON, there will be times when entities might either "snap to themselves" or to some other "undesired point" because of the running OSnap override. If that happens, just toggle the OSnaps OFF. Turn them back on when the procedure at hand is completed.

**OK, DRAW the WINDOW.** You may wish to remove the drawing example from your text for easier coordination with the written instructions.

1. **Line** command: Draw window opening, 3'-4"W × 4'-0"H.

   Snap ON, Ortho ON, "type and drag" 3'4[enter] 4'[enter] 3'4[enter] 4'[enter] [enter]

   Turn Snap (not OSnap) OFF. For reasons to be explained later, it was important to begin on a grid-related Snap, but now that we have the Running OSnaps at our command, we won't be needing the "regular Snap" for a while.

   **Zoom Window** for bigger view.

   You will need to Zoom in and out many times during this drawing process—experiment with **Zoom Window, Previous, All,** and **Realtime.**

2. **Line** command: Draw Lintel (top) and Sill (bottom) 4'-0"W × 4"H.

   Notice that the Endpoint OSnap appears (hopefully) when you select your first point—and final point—at the corners of the window block.

3. **Offset** command: Offset 2" Jambs *to outside* of window opening.

4. **Line** command: Draw middle of frame **Mid**point of top to **Mid**point of bottom.

   The triangle symbol will let you know that the Running OSnaps have done their job. Cool, huh? The OSnaps will put you **EXACTLY** where you want to be.

5. **Offset** 1½" window frames from jambs, sill, lintel, and centerline.

6. **Fillet** "outside corners" of frames. **Zoom** in and out as required.

   For 90° corners—verify Fillet Radius is 0".

   You cannot Fillet the corners at the centerline—you will lose line segments that you want to keep since the Fillet command makes just "one" corner. If you accidentally make this mistake, slowly **UNDO** back to the appropriate view.

7. **TRIM** the frame corners at the centerline. Zoom in and out as required.

READ THE COMMAND LINE FOR THIS OPERATION—THE PROCESS SEEMS BACKWARD TO MANY STUDENTS: The *first* set of objects you select are the cutting edges (the knives). You pick as many "cutters" as you wish, then ENTER. The *last* object(s) you select are the "guys that are leaving town."

---

## SHORTCUT FOR TRIM AND EXTEND: A QUICK RIGHT CLICK

A shortcut for **Trim** (and **Extend**)—call up the Trim command and when prompted for "cutting edges," simply hit the **Enter** key (or **Right Click**) with the cursor free-floating on your drawing screen. This turns *all* lines on your drawing into cutting edges, so you can then begin selecting the actual objects to Trim … as many as you like in one step … Zooming in and out, panning, scrolling your merry way around the screen.

And wait—it gets even better! When you get to the **Extend** command, the same trick will work when selecting so-called "boundaries."

---

Take a minute to think about the differences between **Fillet** and **Trim.** A Fillet makes a single corner out of two lines (intersecting or not). The Trim command (as we are going to use it—other options are possible) requires that lines be "crossing," and only removes segments defined by the "cutting" edges. You can only Trim part of an entity that is "hanging over" another entity. If entities are merely "touching," then Trim won't work … you'll have to **Erase** that puppy. You can make "single corners" with either the Fillet or Trim command.

8. **Line** command: using **Mid**point OSnaps, draw centerlines of muntins (dividers).

9. **Offset** muntins from each side of each centerline. The muntins are $3/4$" wide, so **Offset** half, or $3/8$" each side of centerlines.

10. **Erase** centerlines of muntins—they were just "layout lines."

11. **Trim** muntin intersections—eight locations.

12. **Line** command: Draw the Keystone.

    Toggle OSnap OFF

    Toggle Snap ON (the existing setting is 2")

    Reset Grid to 2" (Tools/Drawing Aids) Toggle Grid ON

    Toggle Ortho OFF—**You cannot draw the diagonal lines if Ortho is ON.**

    Using Snap and the Grid as a guide—draw the Keystone "counting dots," Snapping to Grid.

    **Trim** lintel where it crosses through the Keystone.

**SAVE the drawing…you should be saving at regular intervals while your work is in progress.**

---

## PLOT THE WINDOW IN PAPER SPACE

Use the Quick Guide as needed.

**Select a Layout Tab** (bottom of the drawing area) to enter Paper Space.

**Page Setup:** Right click on the Layout Tab for Page Setup (typically appears by default).

| | |
|---|---|
| **Plot Device Tab:** | (at top of dialog box) |
| **Plotter Configuration:** | per your instructor |
| **Plot Style Table:** | monochrome—for this plot, don't worry about editing (Edit) lineweights; we'll let the lineweights "default." |

**Layout Settings Tab:** (at top of dialog box)

**Paper Size:** letter (8.5×11)

**Orientation:** **landscape**

Other settings are probably OK…your instructor will verify.

**OK** Page Setup and you'll see your designated piece of paper in landscape orientation.

Create a **single Viewport** nearly the size of your paper boundaries. (Load Toolbar if necessary.)

**Activate** and **Scale** the Viewport to 1" = 1'-0"; center the Window in the Viewport (Pan).

Deactivate **Viewport**—Return to Paper Space.

**Save** your finished setup.

**Right click** on the Layout Tab and select **Plot**…**OK.** **OK!**

Notice that you have the option to "**Preview**" your plot prior to sending it to the printer. This is a good way to give things a final once over before giving the big OK.

## MAKING A BACKUP DRAWING FILE

### SAVING THE DRAWING FILE TO A DIFFERENT LOCATION

When Wyatt Earp reportedly asked Doc Holliday for some backup at the O.K. Corral, Doc told Wyatt that he'd "be his Huckleberry." It's a good idea to be your own Huckleberry when using AutoCAD, protecting your own back by SAVING your drawings in their most current version in another location…just in case your original drawing file "gunned down."

You have already "named" and **Saved** the WINDOW drawing to your primary drive location (as indicated by your instructor) when you performed the initial **Save As** at the start of this assignment. And you have updated the WINDOW file at that same drive location by performing **Saves** as you drew. Periodically performing **Saves** as you draw is essential to safe file management—get in the habit of doing that or you run the risk of being a horrible warning for your fellow classmates.

### CREATING A BACKUP: TWO METHODS

Read this stuff and try to keep the blood from running out your ears—but don't proceed without your instructor's direction. He or she will have a preferred method for creating backups and, no doubt, have a demonstration for you as well.

Making a Backup using the Save As command:

**Save to your designated hard drive first—before creating/updating a backup.** In most cases, it is best to be "working" on your designated hard drive location, and create/update your backup file at the END of the work session. To create a backup file, you need to **Save** this drawing under its "current name" but to a "different drive." When you want to **Save** a drawing "As" something "different," you must do a **Save As:**

- Insert a diskette into the (A:) drive—get a disk from your instructor (or provide your own).
- From the **File** Pulldown, select **Save As.** The Save As dialog box will appear.
- At the **Save In** box: select drive (A:)
- At the **File Name** box: the correct name (***WIN) should be displayed.
- Pick **Save**…and you have now saved this drawing in a backup location.

**Repeat this operation at the END of each work session.** Having completed the process once, you will be prompted by a note saying that **"file already exists—do you want to replace it?"**

ALMOST always the answer is YES; you want to replace an older version of the drawing (yesterday's version that you saved) with the newer version (the version that includes today's new, added work). But always think carefully before you answer the replacement question. If you get turned around somehow, you run the risk of replacing "new" work with "old" work—in other words, you lose everything you added to the drawing file today.

There are minor annoyances, if not difficulties with most versions of AutoCAD, using the Save As to backup a drawing, unless you perform it at the end of your work session. Specifically, doing a Save As changes the "Current Drive." A second way of making backups might offer a better option—using Windows™ Explorer.

**Dragging and Dropping Backup Files with Windows™ Explorer:**
Another way to make backup files, a way preferred by many for a variety of reasons your instructor may well share, is to use the Windows™ Explorer and simply "drag and drop" a copy of the original file from one drive to the other. Again, verify the preferred method with your instructor.

You may perform this while "in" AutoCAD and it's really slick...

- **Right Click** on the **START** button at the bottom left of your screen.
- Select **Explore.**
- **Scroll** to your primary drive location in the Folders column on the left.
- **Click** to open the files on your primary drive location.
- Locate your target drawing in the files on the right. It will appear with a .dwg extension designating it as a drawing file.
- Select/highlight the target file.
- **Holding down** the left mouse button, **drag** the file over the drive destination for your backup file (typically the (A:) drive).
- **Release** the mouse button and the file will copy to the backup drive.

Why the Windows™ Explorer for the Backup files? When you perform a Save As while in AutoCAD, you change what we call the Current or "Working" Drive. If you **Create a New Drawing** and do a Save As to a designated drive location, that location becomes the default location for subsequent Saves. That's a good thing.

And likewise, when you **Open** an existing drawing file from a designated drive location, that drive defaults as the Current Drive and any Saves you perform updates the drawing at that drive location.

Typically, if you do a Save As to a different drive location—for instance, saving a backup file to the Floppy A: drive—then your Floppy Drive becomes the Current Drive and your update Saves will start to land on that teeny-weeny disk, which sooner or later will run out of space. Also, Saves to the Floppy Drive are slower.

So we normally want our Current Drive, the Drive on which we are "working" to be one of the big boys—in some cases a server, in some cases the C: drive, in some cases a so-called Jump or Flash Drive.

By saving backups using Windows™ Explorer, you avoid the confusion and frustration of changing Current Drives—using Explorer for your backups will produce no hidden consequences on your Current or Working Drive.

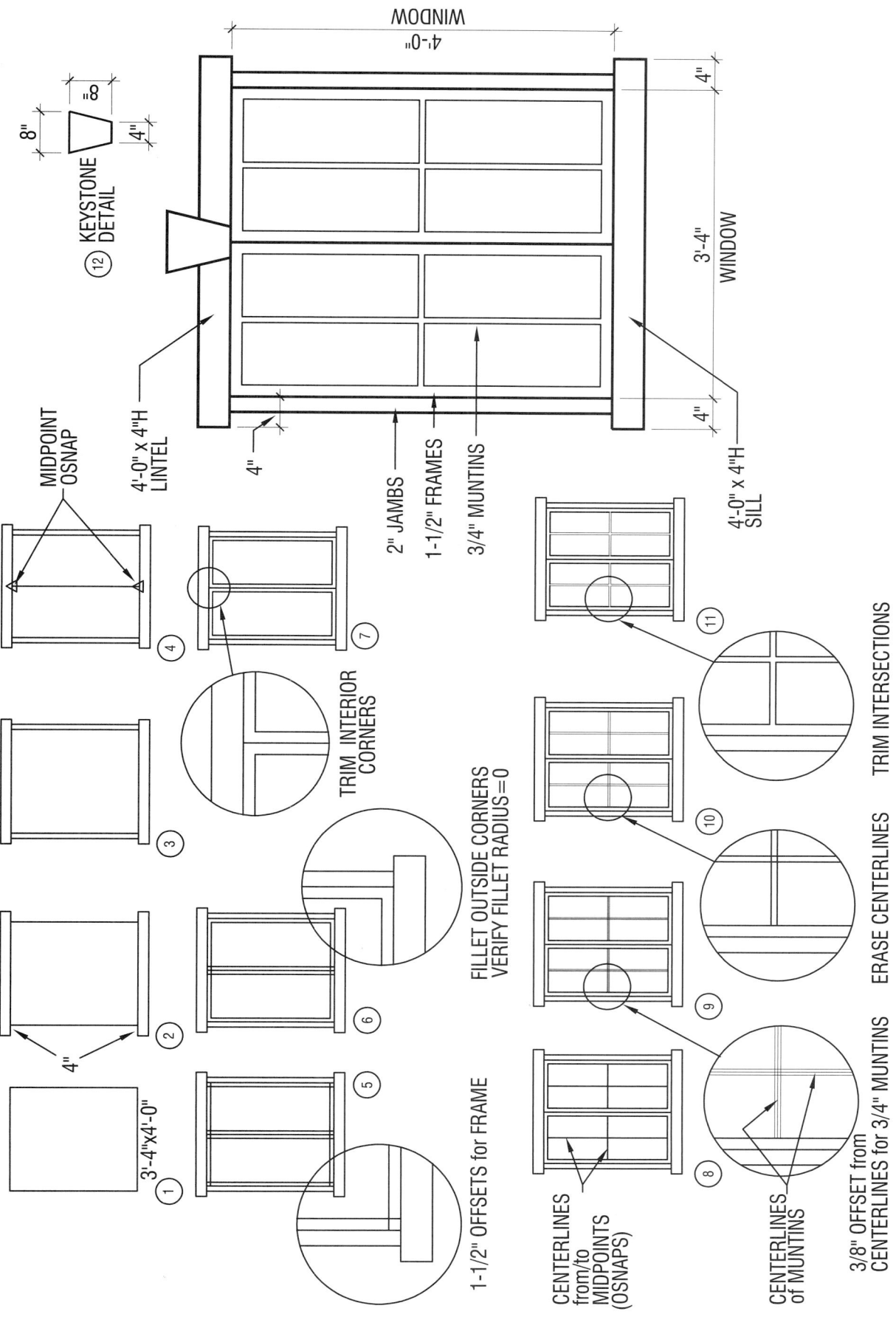

KEYSTONE DETAIL

8"
8"
4"

12 KEYSTONE DETAIL

WINDOW
4'-0"

4'-0" x 4"H LINTEL

4"

2" JAMBS
1-1/2" FRAMES
3/4" MUNTINS

MIDPOINT OSNAP

4"
3'-4" WINDOW
4"

4'-0" x 4"H SILL

1
3'-4"x4'-0"

4"

2

3

4

5

6
1-1/2" OFFSETS for FRAME

7
TRIM INTERIOR CORNERS

FILLET OUTSIDE CORNERS
VERIFY FILLET RADIUS=0

8
CENTERLINES from/to MIDPOINTS (OSNAPS)

9

10

11
TRIM INTERSECTIONS

ERASE CENTERLINES

CENTERLINES of MUNTINS

3/8" OFFSET from CENTERLINES for 3/4" MUNTINS

WINDOW

#2

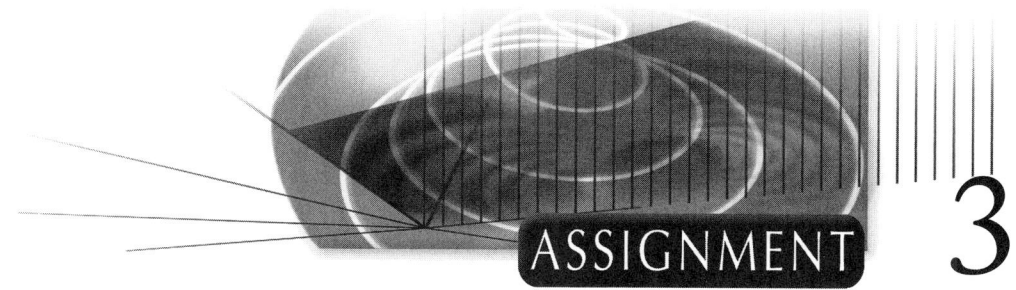

# Chair and Grouping

## COMPETENCIES/LEARNING OUTCOMES

Upon successful completion of the **Chair and Grouping**, the student will have used and begun to master the following settings, operations, and drawing commands.

**Settings:** Units, Limits, Snap, Grid, Polar Tracking, Ortho, OSnap

**Operations:** Open/Exit AutoCAD; Access Menus; Enter Commands; Startup Wizard; Open; Save As and Save drawing files; Creating/Saving backup files.

**Commands:**

    Line

    Offset

    Fillet (0)

    Erase: Select, Window, Crossing Window

    Undo

    Zoom Window, Previous, All, Realtime

    Pan

    Save

    Save As

    Plot in Paper Space

    Trim

    OSnap

**(new)**   **Fillet with Radius (R) setting**

    **Grips**

    **Circle**

    **Copy and Multiple Copy**

    **Mirror**

## PROCEDURE

Working alone and with classmates, incorporate information from lectures, demonstrations, and corresponding information in your text to complete this drawing. In this, your third drawing, you will continue to reinforce lessons/commands encountered in your previous drawings, the Panel Door and Window. Look for similarities in applying the various commands as well as between the creation of this drawing and your earlier assignments. Congratulate yourself on starting to find your way around the AutoCAD screen and for beginning to bring fundamental commands into your control. As always...

**RELAX, EXPLORE, ASK QUESTIONS,**
**WORK WITH YOUR FELLOW CLASSMATES.**

## OK! ANOTHER DRAWING!

### CHAIR AND GROUPING

**Setup:**

If you are in the middle of a work session and finish a drawing, **Save** it one last time, then, from **File** Pulldown, select **Close.** To then begin a new drawing, from the **File** Pulldown, select **New** (or select the appropriate icon). The **Create New Drawing** dialog box will appear and off you go! This stuff should be starting to sound and look familiar.

Pick   **Use a Wizard**

Pick   **Quick Setup**

**Units:** **Architectural** ... **Next**

**Area:**   Width = 22', Length = 17'—make sure you include the "foot mark"

**Finish**

**Zoom All** to ensure that you've captured your entire screen

Pick **Save As** (from **File** Pulldown)

**Save In:** Locate your designated drive—verify with your instructor.

**File Name:** your three initialsCHR—ex: mrbCHR

Pick   **Save**

Remember to **SAVE periodically while working** and **SAVE before you Exit** AutoCAD

**Other Settings:**

**Tools/Drafting Settings: Snap and Grid** tab: Grid 4"—we won't be needing a Snap for this drawing.

**Object Snap** tab:  verify **OSnap** Settings:

Endpoint, Midpoint, Center, Intersection, Perpendicular ON

Other OSnap options OFF.

Always verify these settings with each new drawing!

Verify **Fillet Radius = 0"** Always verify Fillet Radius with each new drawing!

**Toggles:  Snap** OFF; **Grid** ON; **Ortho** ON; **OSnap** ON

You may change these settings and/or turn them on and off as required while drawing.

Notice that we won't be using the Snap setting for this drawing—we will rely on OSnaps instead. That means that the Grid will be used for visual reference only, giving us an idea of "where" we are on our paper. As AutoCAD will let us draw "off" the paper, and knowing that we can also **Move** objects (a command that you will learn later), the relationship between our paper and drawing is not extremely critical at this point.

**DRAW the CHAIR and GROUPING.** You may wish to remove the drawing example from your text for easier coordination with the written instructions.

1. **Line** command: Block out Chair, 2'-8" × 2'-8".

Ortho ON, "drag and type" 2'8[enter] 2'8[enter] 2'8[enter] 2'8[enter] [enter]

### THREE SHORTCUTS

A shortcut: To "close" a shape you're drawing, rather than drawing the final line, type the letter "C" and enter. Presto!

Another shortcut (if Shortcut menu option is Off): Right clicking after a completed command will automatically recall or repeat the previous command—you don't have to keep going back to the Toolbar or Pulldown to repeat. But you must do the extra right click immediately—can't do ANYTHING in between.

And yet another shortcut: For most commands, right clicking on the mouse, hitting either of the enter keys on the keyboard, OR hitting the Space Bar on the keyboard produce the same results

**Zoom Window** "in" for "bigger view."

Again, **Zoom, Pan, Scroll** to aid you in the drawing process.

2. **Offset** sides and fronts of arms, 6" from back, 4" from sides and front.

3. **Trim** the arms, using the **Trim** shortcut—[enter] at the Cutting Edge prompt to turn all edges into cutting edges.

---

## UNDO AND UNDOING SELECTION ENTITIES

If you make a mistake while drawing or selecting, just **Undo** slowly back to familiar territory. If you want to recapture something **Undone**, then **Redo. Undo** is a valuable command, but use it carefully … for AutoCAD Releases *prior* to 2004, you can only **Redo** your "*last* Undo."

For the selecting portion of many commands, if you inadvertently select an entity, type **U** and Enter and AutoCAD will "deselect" the last entity and keep you in the command.

---

4. **Offset** top of Chair back (4").

5. **Offset** sloped portion of back (4"). This portion appears "foreshortened."

6. **Trim** lower part of back between arms.

   **Fillet** "rounded" shapes at the front corners of the arms and cushion.

   Reset the **Fillet** Radius:

   Select **Fillet**

   Type **R** and Enter to access radius option

   Type $^3/_4$ and Enter the value

Depending on your software version, you may have to now "reselect" the Fillet command—check out the new Radius setting … AutoCAD will now default to $^3/_4$" Radius Fillets until you tell it differently. This means that if down the line you need "squared" corners from the Fillet command, you will need to set the Radius to 0".

   **Fillet** the front corners of the cushion first.

   **Fillet** the arms second. Oops! Lost part of the cushion, didn't you!

Had you reversed the order of the Filleting, this would not have happened. But the existing circumstance allows you to try a new tool—**Grips.**

**Grips** provide a quick way to manipulate drawing entities: lines, circles, etc., in a variety of ways, including extending, shortening, and moving the entity. To access **Grips**, pick/select an entity—a line, for example—without entering a command. Just "free pick" on the entity. If you select a line, the line will highlight and **Grip** boxes will appear at the Endpoints and Midpoint. By left clicking on a Grip, it "turns hot." This will allow you to "pick up the end of a line" and reposition it (left click). For accuracy, use OSnap targets for the new location, or drag and type distances. Often using the perpendicular OSnap for the target area will result in similarities to the **Trim** and **Extend** (next unit) commands. Your instructor will demonstrate.

Sometimes when Running OSnaps are ON, the entity will snap back to itself or to an undesired location—in those cases, turn the OSnap toggle OFF. Don't forget, however, to turn OSnaps back ON when you continue drawing.

   So … to close the gaps at the cushion corners using **Grips** …

   **Zoom** in at a front corner of the Chair.

   With **no command** at the prompt line, **select the outside line** of the arm. The line will highlight and **GRIP** boxes will appear.

   **Left click** on the Endpoint **GRIP** to "pick it up"—it will change color—turn "hot."

**Left click** on your target to put the Endpoint "back down," in effect stretching the line to the end of the rounded cushion corner. Your running OSnap will find the point for you. COOL, HUH? Remember these drawing aids; rely on GRIPS for quick fixes as you draw.

7. Draw four buttons:

   **Offset** 6" from insides of each arm and the front and back seat cushion.

   The resulting intersections will give you OSnap targets for each button.

   Erase these layout lines when you've located all four buttons.

   **Circle** command:

   Draw only ONE button—we'll use the **Copy** command for the remaining three.

   Pick **Circle**

   Select **Center**—Intersection OSnap at one location defined by layout lines.

   Radius = $^1/_2$ (Button Diameter = 1" so Radius = $^1/_2$"—right?) Enter.

---

### DRAWING CIRCLES

AutoCAD typically asks for the Radius of Circles as its default, so if you know the Diameter, then divide it by 2 for the R value. If you wish to use Diameter, type **D** and **Enter**—then enter the value for the Diameter.

---

   **Copy (Multiple)** the button to the other three locations.

   Zoom in at the Chair cushion for a "bigger view." During the **Copy** command, AutoCAD will allow you to Zoom and Pan as needed.

   Select the **Copy** command—**Modify** Pulldown or icon

   Select the button (object(s) to Copy)

   **Enter** to tell AutoCAD that you are done selecting objects to **Copy.**

   Type **M**, then **Enter,** for **Multiple** copies (Release 2005 does this by default).

   **Select the intersection** of the existing button's layout lines for **Base Point.** Let your OSnap find this point—you may need to Zoom in.

   **Select** remaining three intersections described by your layout lines for **Displacement** points.

   **Erase** the layout lines when the four buttons have been located.

   **Copy** command—AutoCAD will allow you to copy as many entities as you wish—so you must tell it when you're done selecting objects to copy by right clicking or hitting the Enter key. This is true for most all "selecting" operations—AutoCAD will keep asking you to "select objects" until you "tell it" you're done doing so.

   Think of the **Basepoint** and **Displacement** points in terms of "from here"—"to there." These points are relative. Neither the Base nor Displacement points need be "on" the object. Also, you can "drag, type, and enter" displacement distances similar to the manner in which you type distances when drawing lines, Copying entities exact distances above, below, to the right, to the left, etc., of the existing entity. Finally, you may also use OSnaps to precisely locate Base and or Displacement points.

   You will find that selecting entities for other upcoming commands—the **Mirror, Move,** and **Array** commands—work in similar fashion.

8. **Copy** your finished Chair 24" to its right side.

   Select the **Copy** command.

   **Select objects**—Capture the entire chair in a Pick Window.

   **Enter** to tell AutoCAD that you are done selecting objects.

**Base** point—select the upper left corner of the chair.

**Displacement**—Ortho ON—drag to the right and type 4'8 (4'-8"), **Enter.** BINGO!

Note that for a space of 2'-0" between the chairs, each entity of the original had to move (be displaced) 4'-8"—that's 2'-8" for the chair itself *plus* the 2'-0" clearance.

9. **Mirror** the pair of Chairs to create a grouping of four.

As part of the **Mirror** command, you will be asked to draw a "mirror line." A mirror line is an imaginary line splitting the distance between reflected images. Picture a wet smear of ink on a piece of paper. You fold the paper in half and the ink blot transfers as a "mirrored image" to a "mirrored" location. The fold or crease in the paper may be viewed as the mirror line.

Draw the *imaginary* Mirror Line as you would any other line—by picking a starting point and an ending point. You will see it dragging on your screen, but it will disappear after you draw it. For accuracy, we often need to give ourselves a layout line—or find convenient OSnap locations for the start and end of the mirror line. In most cases, you will want Ortho ON when drawing the mirror line. AutoCAD extends the mirror line to infinity—you need not draw it "all the way across" the entities you wish to **Mirror.**

To **Mirror** the two chairs 5'-0" apart, we need a mirror line (fold in our paper) 2'-6" from the front of the chairs (half the total distance).

**Offset** a short layout line—the front of one of the chairs will do—at a distance of 2'-6" for reference. This will provide a reference for upcoming the mirror line.

**Select** the **Mirror** command—**Modify** Pulldown or icon

**Select Objects**—capture both Chairs in a Window.

**Enter** to tell AutoCAD that you're done selecting objects—sound familiar?

**Mirror line**—draw the Mirror Line from Endpoint to Endpoint of the layout line.

**Delete Source**?—type/select N (for No) and then Enter.

   QUADRUPLE BINGO!—FOUR CHAIRS from ONE!

**Erase** the "layout line."

**Save**—you have been Saving periodically as you drew—right?

---

## PLOT THE CHAIRS IN PAPER SPACE
Use the **Quick Guide** as needed.

**Select a Layout Tab** (bottom of drawing area) to enter Paper Space.

| | |
|---|---|
| **Page Setup:** | Right click on the Layout Tab for Page Setup. |
| **Plot Device Tab:** | (at top of dialog box) |
| **Plotter Configuration:** | per your instructor |
| **Plot Style Table:** | monochrome—for this plot, don't worry about editing (Edit) lineweights; we'll let the lineweights "default." |
| **Layout Settings Tab:** | (at top of dialog box) |
| **Paper Size:** | letter (8.5 × 11) |
| **Orientation:** | **landscape** |

Other settings are probably OK…your instructor will verify.

**OK** Page Setup and you'll see your designated piece of paper in landscape orientation.

Create a **single Viewport** nearly the size of your paper boundaries. (Load Toolbar if necessary.)

**Activate** and **Scale** the Viewport to $1/2$" = 1'-0"; center the four chairs in the Viewport (Pan).

Deactivate Viewport—Return to Paper Space.

**Save** your finished setup.

**Right click** on the Layout Tab and select **Plot**... try a Preview... **OK.**       **OK!**

Notice that you have the option to "**Preview**" your plot prior to sending it to the printer. This is a good way to give things a final once-over before giving the big OK.

### SAVE this drawing before you exit.

**Save** to the drive location designated by your instructor and create a Backup file on the Floppy Drive (A:) either using Windows™ Explorer or by doing a **Save As**... per your instructor's direction.

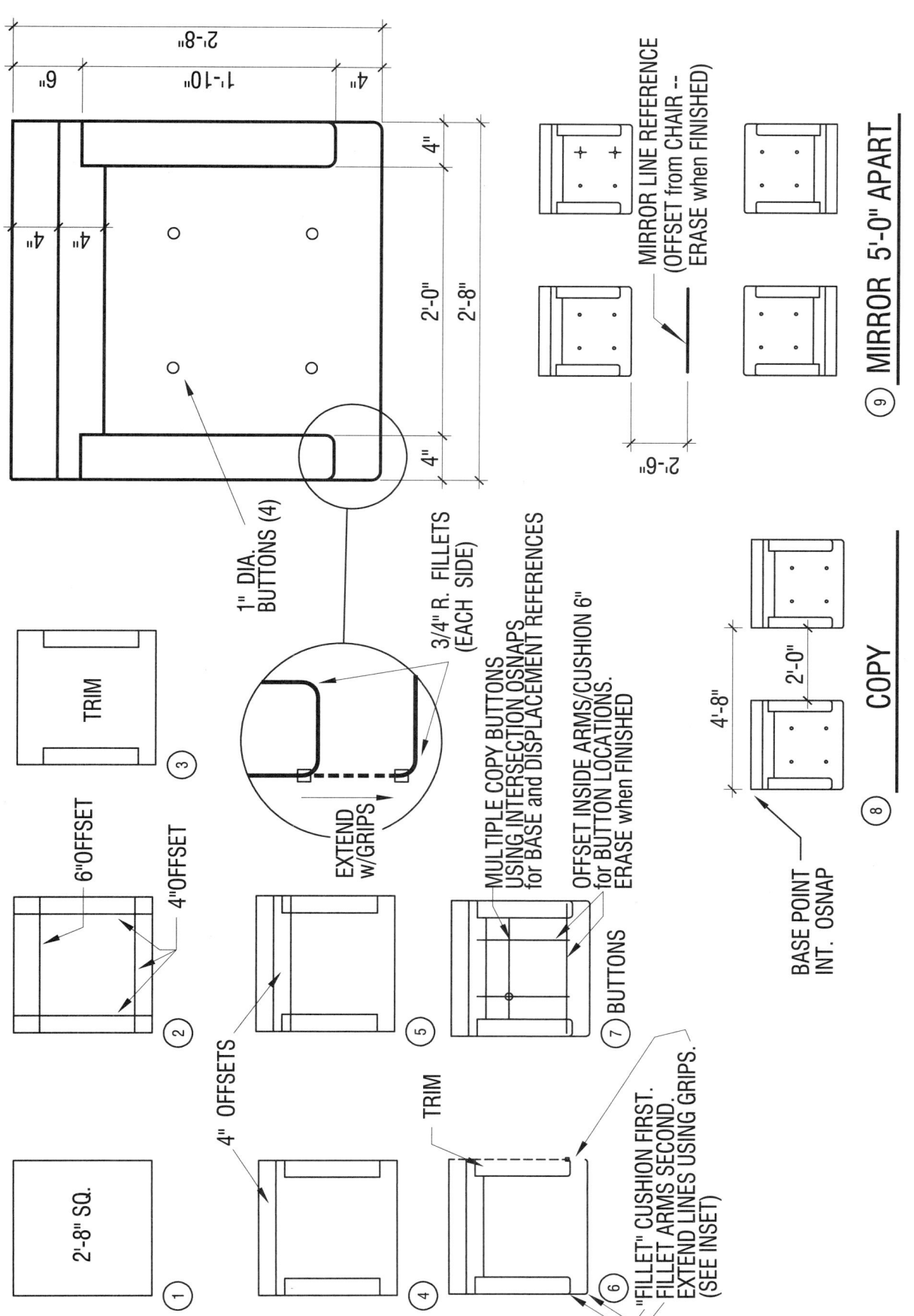

2'-8"

1'-10"

6"

4"

4"

4"

4"

2'-0"

2'-8"

4"

1" DIA. BUTTONS (4)

3/4"R. FILLETS (EACH SIDE)

EXTEND w/GRIPS

TRIM

③

6"OFFSET

4"OFFSET

②

4" OFFSETS

2'-8" SQ.

①

TRIM

④

"FILLET" CUSHION FIRST. FILLET ARMS SECOND. EXTEND LINES USING GRIPS. (SEE INSET)

⑥

MULTIPLE COPY BUTTONS USING INTERSECTION OSNAPS for BASE and DISPLACEMENT REFERENCES

OFFSET INSIDE ARMS/CUSHION 6" for BUTTON LOCATIONS. ERASE when FINISHED

⑤

⑦ BUTTONS

MIRROR LINE REFERENCE (OFFSET from CHAIR -- ERASE when FINISHED)

2'-6"

⑨ MIRROR 5'-0" APART

BASE POINT INT. OSNAP

4'-8"

2'-0"

⑧ COPY

CHAIR GROUPING

#3

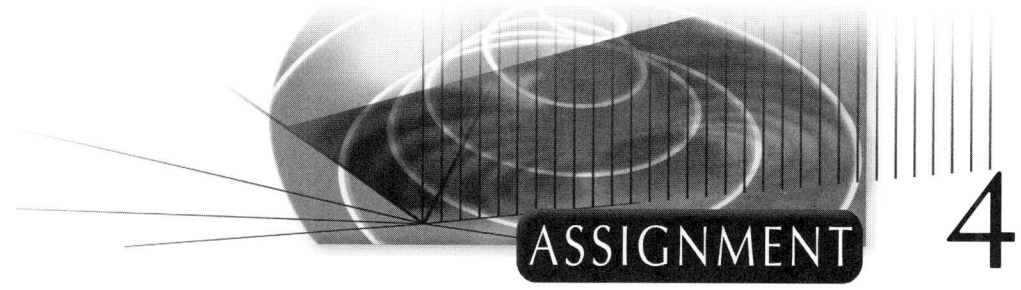

# Floor Plan: Block Out

## COMPETENCIES/LEARNING OUTCOMES

Upon successful completion of the **Floor Plan: Block Out,** the student will have used and begun to master the following settings, operations, and drawing commands.

**Settings:**   Units, Limits, Snap, Grid, Polar Tracking, Ortho, OSnap, **Layers, Linetypes**

**Operations:**   Open/Exit AutoCAD; Access Menus; Enter Commands; Startup Wizard; Open; Save As and Save drawing files; Creating/Saving backup files.

**Commands:**

Line

Offset

Fillet (0)

Erase: Select, Window, Crossing Window

Undo

Zoom Window, Previous, All, Realtime

Pan

Save

Save As

Plot in Paper Space

Trim

OSnap

Fillet with Radius (R) setting

Grips

Circle

Copy and Multiple Copy

Mirror

**(new)**   **Layer Properties**

**Linetype Load/Selection**

**Linetype Scale**

**Extend**

**Properties (Modify and Match)**

## PROCEDURE

Working alone and with classmates, incorporate information from lectures, demonstrations, and corresponding information in your text to complete this drawing. In this, your fourth drawing, you will continue to reinforce lessons/commands encountered in your previous work.

This drawing will eventually evolve into your "mid-term Opus." We will return to it over the next eight assignments, adding detail, dimensions, text, and "**Inserting**" previous and subsequent drawings.

As you continue to build your expertise, look for similarities in applying the various commands as well as between the creation of this drawing and your earlier assignments. Your "instruction" will become less and less descriptive as you begin to master the commands and rely on your own developing expertise. Look for strategies in the layout and operations, think of different ways—and the best ways—of accomplishing specific tasks. As always...

**RELAX, EXPLORE, ASK QUESTIONS,**
**WORK WITH YOUR FELLOW CLASSMATES.**

## LINETYPES AND LAYERS

First, some words from your sponsor about Linetypes and Layers ... an overview to accompany your instructor's demonstrations.

### Linetype and Linetype Scale

We use a variety of Linetypes when completing an architectural drawing including continuous (solid) lines, hidden lines (a "dashed" line used to show objects hidden from our view), and centerlines (to indicate an axis of symmetry or to locate "centers" of entities). AutoCAD has an extensive menu of Linetypes for us to use, but until we "load" those types into our drawing file, we typically can only draw continuous lines.

If you need lines other than continuous lines, you must **Load** the different Linetypes **each time you begin a NEW drawing.** Once loaded into a drawing file, they remain—you do not have to "reload" them each time you work on that particular drawing. You may select specific Linetypes to load, or you may use the **Select All** option to load "all" the Linetypes AutoCAD affords. For architectural drawings, we most typically need continuous, hidden, center, and phantom lines.

Linetypes may be loaded using the **Linetype** command from the **Format** Pulldown (other options exist) and then employing the **Linetype Manager** dialog box. Pick **Load** and select types you need. Pick **Show Details** button (if not displayed) to set Linetype Scale—more on these things later.

### Layers

As part of this exercise you will also be asked to create **Layers.** Later, as part of subsequent exercises, you will be asked to allocate certain portions of this drawing to those **Layers.** "Layering" is important in architectural drawings—among other available operations embedded in Layer Control, we can manipulate our final Plot by turning various Layers "ON" or "OFF." For instance, we will draw dimensions on a Dimension Layer, and then if we want a print of the Floor Plan WITH dimensions, we run a Plot with the Dimension Layer turned "ON," and if we want a Plot of the Floor Plan WITHOUT the dimensions, we run a Plot with the Dimension Layer turned "OFF."

Layer creation and management then becomes an integral part of organizing your drawing file. Each office may have slightly different Layering Formats, but the concepts you will learn here are basic to all operations.

While additional, more sophisticated layer-related operations are available through AutoCAD, remember we are "beginning" here—so focus on the methods of creating NEW Layers, selecting their color and Linetype, designating the Current Layer, and assigning drawing entities to their proper Layer.

The **Layer Properties Manager** dialog box can be accessed through the **Layer** command in **Format** Pulldown or by selecting the **Layer Properties Manager** icon in the object **Layers Toolbar.**

**Assignments for Linetypes and Layers:**
Assigning different Linetypes and assigning entities to their proper Layer—plumbing fixtures on the Plumbing Layer, doors on the Door Layer, exterior walls on the Exterior Wall layer, etc.—can be achieved through several different routes. Quite typically a line will be laid in as a continuous line, and then its properties changed to a centerline or hidden line. In similar fashion, an entity may "originate" on your screen on a mismatched Layer, say through an Offset command, and then you will move it to the proper Layer, again by changing, or **Modifying,** its properties. The **Properties Palette** will come into play, accessed from the **Modify** Pulldown or the **Properties** icon in the Standard Toolbar. Shortcuts for making these changes are also available and will be discussed by your instructor and in the following information.

> **WHEW! Enough of that stuff, let's get things ready to draw.**

## FLOOR PLAN: BLOCK OUT

### NOW FOR THE BIG TIME—A FLOOR PLAN OF A *GEORGIAN HOUSE*
You will begin a new drawing for the Floor Plan, so on a "clean screen" from the **Create New Drawing** dialog box…

Pick     **Use a Wizard**

**Quick Setup**

**Units: Architectural**…**Next**

**Area:** Width = 144', Length = 96' (be sure to include the "foot mark")

**Finish**

**Zoom All** to ensure that you've captured your entire screen.

Pick **Save As** (from **File** Pulldown) to rename this drawing.

**Save In:** Locate your designated drive—verify with your instructor.

**File Name:** your three initialsGRG—ex: mrbGRG

> Remember to **SAVE periodically while working** and **SAVE before you Exit** AutoCAD

**Other Settings:**

**Tools/Drafting Settings: Snap and Grid** tab: Grid 4'—no Snap setting needed

> **Object Snap** tab: verify OSnap Settings—
>> Endpoint, Midpoint, Center, Intersection, Perpendicular ON
>> Other OSnap options OFF

Always verify these settings with each new drawing!

Verify **Fillet Radius = 0"** Always verify Fillet Radius with each new drawing!

**Toggles: Snap** OFF; **Grid** ON; **Ortho** ON; **OSnap** ON

> You may change these settings and/or turn them on and off as required while drawing.

Ok—that part of a drawing setup should be approaching "routine." For this drawing you also need to Load Linetypes and set a number of Layers. Once you've accomplished this—the settings remain a permanent part of THIS drawing file. Other New drawings down the line will require these—and additional—operations. While AutoCAD has more than one way to transfer settings of this nature from an old drawing to a new drawing, you'll be asked to repeat Setups for your new drawings. Tough love here—as a beginner, the practice serves you well.

## LOAD LINETYPES, SET LINETYPE SCALE, AND SET LAYERS

**To Load Linetypes and Set Linetype Scale:**
From the **Format** Pulldown menu…

Pick **Linetype** to access **Linetype Manager** dialog box.

Pick **Load** to access the **Load Linetype** dialog box.

"Scroll" to **Center,** then HOLDING DOWN THE CONTROL (**CTRL**) KEY,

Pick **Center, Center2, Centerx2**…Release the **CTRL** key

"Scroll" to **Hidden,** then HOLDING DOWN THE CTRL KEY,

Pick **Hidden, Hidden2, and Hiddenx2.**

Pick **OK.**

**Still in the Linetype Manager Dialog box …**

Pick **Show Details** (if needed) from the upper right screen to set **Linetype Scale.**

(if **Hide Details** is displayed at the upper right screen, then this step is not necessary)

Set **global scale factor** to **24.**

Set **current object scale** to **1.**

Deselect—"uncheck"—the box at **"use paper space units for scaling"**

**OK**…returns you to your drawing screen.

You have now loaded the Linetypes listed above.

Note: If you wanted to load ALL the Linetypes AutoCAD has to offer, instead of picking each one individually, you may:

**Right click** your mouse to access the small **Select/Clear** box.

**Left click "Select All."**

**Left click "OK"** and set Global Linetype Scale as above.

This strategy is simpler…but not recommended.

---

### A NOTE ABOUT LINETYPE SCALE:

Adjusting the Linetype Scale "boosts" the "gaps" in Hidden and Centerlines so they "read" at our small Architectural Scales. If hidden, center, or other types of "broken" lines are reading as "continuous lines" on your drawing, chances are you need to adjust the Linetype Scale. The little gaps are there—they are just too small for you to see. You entered a Linetype Scale of **24** for this drawing. The Linetype Scale is often set for "half" the scale factor for the final plot. If we wish to plot our floor plan at $^1/_4" = 1'\text{-}0"$ scale, that's $^1/_{48}$ actual size—a scale factor of 48. So if we wish our Linetype Scale to be half of that, we wind up with 24 (*Supplement 2*).

---

**Set Layers**
From the **Format** Pulldown select **Layer**…or…pick the **Layer Icon** to access the **Layer Properties Manager** dialog box.

Notice AutoCAD defaults to Layer 0—you drew the Panel Door, Window, and Chairs on Layer 0… probably didn't know it, but you have already been using Layers! AutoCAD demands that you keep Layer 0 in the mix—you can't delete it. But you can add new layers…

Create new Layers with these respective Names, Colors, and Linetypes:

| | | | |
|---|---|---|---|
| WALLS_EXTERIOR | WHITE | CONTINUOUS | |
| WALLS_INTERIOR | YELLOW | CONTINUOUS | |
| WALLS_FURRING | MAGENTA | CONTINUOUS | |
| DOORS | YELLOW | CONTINUOUS | |
| CABINETS | BLUE | CONTINUOUS | |
| PLUMBING | CYAN | CONTINUOUS | |
| ELEVATIONS | WHITE | CONTINUOUS | |
| DIMENSIONS | RED | CONTINUOUS | |
| TEXT | GREEN | CONTINUOUS | |
| FURNITURE | MAGENTA | CONTINUOUS | |
| APPLIANCES | CYAN | CONTINUOUS | |
| VIEWPORT | WHITE | CONTINUOUS | NO PLOT |

**To Set Layer Names:**

Select **New** and "**Layer 1**" will appear in colored highlight…ready to be typed over with a new name, set to a new color, and linetype.

**After you set the desired Name, Color, Linetype, etc.—just Enter and the process repeats for the next New Layer.**

---

**NAMING LAYERS**

Layer names (or drawing file names for that matter) typically can't have colons, periods, or slashes— it's a DOS thing. Though we don't need to enter them so, AutoCAD will display Layers in alphabetical order, so grouping Layer names (Walls_Ext, Walls_Int) will result in displaying all Wall Layers in the same vicinity. Easier to "find" things that way.

---

**To Set Layer Color:**

**Below "Color"** on the control panel you find selection boxes for each Layer. These boxes indicate the Color assigned to each Layer. White(black) is the default. White is our choice for our Exterior Walls (WALL_EXTERIOR), but we wish to assign alternate Colors to other Layers. This will help us "read the drawing" on the screen, and importantly, the different Colors can, if we choose, be used to Plot lines at different widths when we Plot our drawing. You will discover that you can have more than one Color on a Layer—you're just setting the typical Color at this point.

After naming the new Layer, select the **Color** box and assign the appropriate Color. Pick the designated Color from the standard Color bar—Red, Yellow, Green, Cyan Blue, Magenta, White, Gray 8, Gray 9— **Colors 1–9** displayed near the center of the dialog box. **At this time, don't select from the wider range of Colors displayed at the top of the dialog box.**

Click **OK** to return to the previous screen.

**To Set Layer Linetypes:**

The process for assigning Linetypes to a Layer is the same as assigning Colors. Click the default Linetype (Continuous) to access the Linetype Menu *that you have loaded*. If you had not loaded the Linetypes as you did earlier for this drawing, when you access this screen only continuous lines will appear as an option. We are fine with the default settings for now—we want these layers to be drawn with continuous lines so no adjustments are necessary. But go "into" the Linetype screen to have a peek.

**Enter** to repeat the process for setting up another New Layer.

**Setting a Current Layer:**

The Current Layer is the Layer "on which you are drawing," i.e., if we designate the Plumbing Layer as Current, then lines we draw on our screen will automatically be assigned to that Layer. By default, in your previous drawings, the "Current Layer" has been Layer 0. Entities Offset, Copied, etc., will retain the Properties of the source Layer. You may need to reassign such entities to their proper Layer after the initial operation is complete.

Pick the **WALL_EXTERIOR** Layer and then select **Current.** This means that when you begin drawing your exterior walls, they will be allocated to this drawing layer.

**Select OK ... be sure to "OK" your way out of the Layer Properties Screen.**

Be sure you **"OK"**—if you "Cancel" your way out—then all your settings will be canceled as well!

Once you're back to the drawing screen ... **Save!** Don't want to risk having to do that again, do you?

Check the Layer display at your Layers Toolbar to confirm the **Current Layer,** check the Properties Toolbar to confirm **Color (ByLayer)** and **Linetype (ByLayer).** You may easily change the Current Layer by using the fly out arrow at the display window, and selecting a new "Current Layer"—or you may return to the Layer Properties Manager dialog box to change the Current Layer.

Again, with the help of your instructor if needed, **verify that Color and Linetype assignments read ByLayer** at the Properties Toolbar. If these settings read otherwise, then Color and/or Linetype assignments for Each Layer will be overridden.

## NOW YOU CAN DRAW ... FINALLY

You may wish to remove the drawing example from your text for easier coordination with the written instructions.

One strategy for drawing a building like this (it's bilaterally symmetrical) is to draw a portion of it and then **Mirror** the remaining parts ... so begin by drawing **one quarter of the BUILDING**—what we will call a "quarter section."

If you've performed a Zoom ALL you are seeing a 48"(4') Grid on your screen. Again we are using the Grid as a visual reference to "see" our paper—nothing more. Snap is OFF, and for that matter, you may prefer to turn your Grid OFF for "clearer" drawing.

A reminder: **OSnap ON, Ortho ON.**

**Zoom** the bottom right quarter of the screen to begin—you will probably need to adjust the Zoom several times for drawing ease.

1. **Line command:** Pick a starting point, "type and drag" with Ortho ON ...

   Draw the outline for the bottom right quarter of the house. The "starting point" is no big deal—we can move the drawing around later to position it better on the sheet.

2. Draw two intersecting Lines that OSnap from the endpoints of the quarter section. These lines show up as White, Continuous lines—right? Those are the default settings "ByLayer" for the WALLS_EXTERIOR Layer, your Current Layer.

   To **modify** (change) these "continuous" lines to "centerlines":

   Select **Properties** from the **Modify** Pulldown ... or ... Pick the **Properties** icon.

   The **Properties Palette** will appear.

   "Pick" **the two lines** without entering a command ... like you did to access Grips.

   In **General** ... in the Properties **Palette**

   > Select **Linetype** and then from the fly out, select **Center.**

   > Select **Color** and then, from the fly out, select **Green.**

Selecting the **X** at the top will close the Properties dialog box ... you may "dock" it by selecting the AutoHIDE arrows at the bottom corner ... it'll be waiting for you there.

The **Properties** command is powerful. As you can see from the dialog box you can change **Color, Linetype, and/or Layer** of entities, along with myriad other entity properties using this command. This also means that despite your earlier Layer settings, AutoCAD will allow you to have different Colors and different Linetypes on the same Layer. This is a good thing, if we assign Plotted Lineweights by Color and wish to have different Lineweights plotted on the same Layer. Some prefer other methods, including creating separate Layers for entities of differing Lineweights; Modifying Lineweights directly, independent of Color; and/or assigning Lineweights "ByLayer" instead of "By Color."

---

### SHORTCUT TO MODIFYING LAYER, COLOR, LINETYPE

You may also change Linetype, Color, and/or Layer by directly "picking" an entity(s), and then going to the Layer, Color, and/or Linetype flyout arrows in the **Layers** and **Properties Toolbars** across the top of your screen, selecting the desired change. This is a much simpler and faster way to make a change—but does not give you the range of options offered in the dialog box. Try this—you'll like it. Color and Linetype boxes should read BYLAYER when drawing!

---

### MATCH PROPERTIES: ANOTHER SHORTCUT FOR MODIFYING PROPERTIES

A related command to **Modify/Properties** is the **Match Properties** command—the **paint brush** icon. If you already had a *green centerline*, you could pick the **Match Properties** icon, pick the source (*green centerline*), and then pick the line or lines you wanted to match, or "turn into," green centerlines. A danger to remember is that the changed entities will also be matched to the sources Layer as well (using [S]ettings option will allow matching individual Properties). Experiment with this command; it is easy and powerful. You will come to value it.

---

3. Using **Grips,** stretch the Green Centerlines a bit past the Wall Endpoints.

   **Offset** wall thicknesses 8".

4. **Fillet** wall corners—**verify your Fillet Radius = 0".**

5. **Offset** layout lines for Jambs (sides) of Windows and for window glass.

   Assume glass is at middle of 8" wall—**Offset 4".**

6. **Zoom** in and **Trim** to finish windows.

   • Remember that "right clicking" will turn all edges into cutting edges.

   • Remember that you can **Zoom** and **Pan** while in the **Trim** command.

7. **Extend** the lines of the interior Hall Wall from the Entry corner to the centerline.

   Select **Extend** command from the **Modify** Pulldown or Toolbar icon.

   **Select Boundary Edges,** i.e., "where you want to go" first.

   **Enter** when done selecting Boundary Edges.

   **Select the Object to Extend.** Try to pick the extending line near the end that is to grow.

   **Trim** out the corner at the entry for clean intersections.

---

### EXTEND SHORTCUT

The same shortcut you used to select cutting edges for **Trim** works for selecting Boundary Edges for **Extend.** After selecting the command, "right click" and ALL edges become Boundaries. You may then directly pick the objects to extend. This may or may not prove to be a shortcut—sometimes it's more efficient to select the specific boundary for the extension.

8. **Mirror** the Quarter Section to create half the finished block out.

    **Select Mirror** from the **Modify** Pulldown (or icon).

    **Select Objects:**

- Capture the quarter section in a window.
- Try to avoid capturing the centerline.

    **Enter** to tell AutoCAD you're done selecting objects.

    Draw **Mirror Line,** starting and ending at the Endpoint OSnaps of the centerline.

    **Delete Source?—N ... Enter.**

9. **MIRROR** the Half Section to create the finished block out.

    Repeat the process above, capturing the half section for the mirror. Use the "other" centerline as your OSnap reference for the mirror line ...

and **BINGADEE BANGADEE BOO—A COMPLETED BLOCK OUT!**

**IS THIS A GREAT COUNTRY OR WHAT!**

10. **Extend** centerlines as shown, using GRIPS.

---

## PLOT THE FLOOR PLAN: BLOCK OUT IN PAPER SPACE

Use the **Quick Guide** for Plotting in Paper Space as needed.

**Select a Layout Tab** to enter Paper Space:

| | |
|---|---|
| **Page Setup:** | Right click on the Layout Tab for Page Setup. |
| **Plot Device Tab:** | (at top of dialog box) |
| **Plotter Configuration:** | per your instructor |
| **Plot Style Table:** | monochrome—for this plot, don't worry about editing (Edit) Lineweights ... we'll let the Lineweights "default." |
| **Layout Settings Tab:** | (at top of dialog box) |
| **Paper Size:** | letter (8.5 × 11) |
| **Orientation:** | **landscape** |

Other settings are probably OK ... your instructor will verify.

**OK** Page Setup and you'll see your designated piece of paper.

Create a **single Viewport** nearly the size of your paper boundaries.

    **Scale** the Viewport to $1/8'' = 1'\text{-}0''$; center the Block Out in the Viewport (Pan).

    (Plans are typically plotted a $1/4'' = 1'\text{-}0''$, "check plots" are often run at smaller scales)

    Place the Viewport on the **VIEWPORT LAYER** (we set that Player "not to plot"—remember?)

**Save** your finished setup.

**Right click** on the Layout Tab and select **Plot.**

**SAVE THIS DRAWING BEFORE YOU EXIT.**

**Save** to the drive location designated by your instructor and create a Backup file on the (A:) Drive using either Windows™ Explorer or by doing a **Save As** ... per your instructor's direction.

**CENTERLINES**

STOP DRAWING
QUARTER SECTION
HERE

START DRAWING
QUARTER SECTION
HERE

16'-8"

4'-4"

4'-8"

3'-4"

3'-4"

4'-8"

19'-4"

3'-4"

4'-8"

5'-4"

2'-8"

USE "O-SNAP" TO
LOCATE END OF LINES

DRAW INTERSECTING LINES
from QUARTER SECTION ENDS.
CHANGE TO CENTERLINES

② 

FILLET CORNERS
SET FILLET @ "0" RADIUS !!

④ 

TRIM WINDOWS

⑥ 

LAY OUT LOWER RIGHT
QUARTER of the PLAN

① 

EXTEND CENTERLINES
USING GRIPS

OFFSET WALLS
THICKNESS = 8"

③ 

OFFSET WINDOW
JAMBS and
GLASS

4'-8"

⑤ 

BOUNDRY

EXTEND WALL

TRIM
INTERSECTIONS

⑦ 

⑨ MIRROR HALF SECTION

REFERENCE for
MIRROR LINE

⑧ MIRROR QUARTER SECTION

USE CENTERLINE as
REFERENCE for
MIRROR LINE
(OSNAP to EACH END)

**GEORGIAN HOUSE: FLOOR PLAN -- BLOCK OUT**

#4

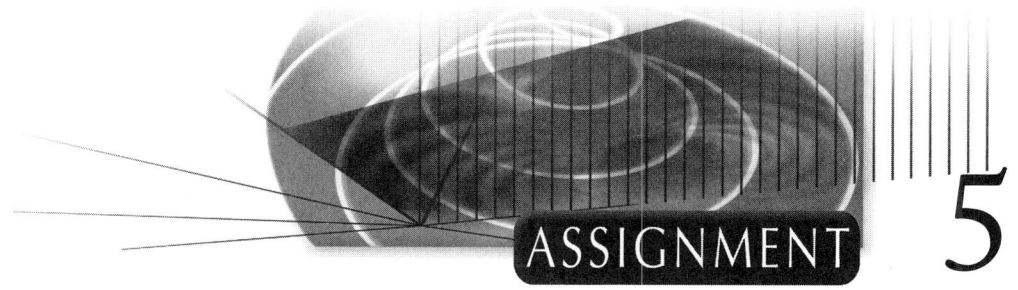

# Floor Plan:
# Interior Features

## COMPETENCIES/LEARNING OUTCOMES

Upon successful completion of the **Floor Plan: Interior Features,** the student will have used and begun to master the following settings, operations, and drawing commands.

**Settings:**     Units, Limits, Snap, Grid, Polar Tracking, Ortho, OSnap, Layers, Linetypes

**Operations:**  Open/Exit AutoCAD; Access Menus; Enter Commands; Startup Wizard; Open; Save As and Save drawing files; Creating/Saving backup files.

**Commands:**

Line

Offset

Fillet (0)

Erase: Select, Window, Crossing Window

Undo

Zoom Window, Previous, All, Realtime

Pan

Save

Save As

Plot in Paper Space

Trim

OSnap

Fillet with Radius (R) setting

Grips

Circle

Copy and Multiple Copy

Mirror

Layer Properties

Linetype Load/Selection

Linetype Scale

Extend

Properties (Modify and Match)

**(new)**  **Arc: SCE (Start, Center, End)**

## PROCEDURE

Working alone and with classmates, incorporate information from lectures, demonstrations, and corresponding information in your text to complete this drawing. In this, your fifth drawing, you will continue to reinforce lessons/commands encountered in your previous work.

As you continue to build your expertise, look for similarities in applying the various commands as well as between the creation of this drawing and your earlier assignments. Look for strategies in the layout and operations, think of different ways—and the best ways—of accomplishing specific tasks. As always…

**RELAX, EXPLORE, ASK QUESTIONS,
WORK WITH YOUR FELLOW CLASSMATES.**

## COLOR-BASED PLOTTING STYLES

### COLOR AND LINEWEIGHT (WIDTH)

When drawing by hand on the drafting board, we draw lines to different widths, for the drawing to "read." You may notice that on your computer screen, all lines (for now) appear to be the same width. We will wait until we Plot to assign different line widths, or **Lineweights,** to our drawings. We can do this in two ways—either by assigning Lineweights to each Layer, in the Layer Properties Manager dialog box, or by assigning different Lineweights to different Colors on our screen. These groups of exercises will choose the later—assigning Lineweights by Color. Your instructor may request you to follow the alternate route.

*Supplement 6: Quick Guide to Plotting in Paper Space* shows Color assignments for various Lineweights for these groups of exercises. Refer to the charts describing Colors and their Lineweights. Understand that these Color assignments are arbitrary and different offices or schools may have different standard Color assignments. Whatever the Color assignments for Lineweights, start "seeing" different Lineweights when you view different Colors on the screen. For example, if Exterior Walls are shown in WHITE, we will tell the computer to Plot all WHITE lines "very wide." In turn, this means that we would NOT want to show "thin" entities like dimension lines as WHITE, but rather as a designated "thin" Color—in our case, RED. You are about to also confirm that you may have more than one Color on a single Layer, meaning that you will be able to Plot different Lineweights within the context of a single Layer. For example, Base Cabinets and Wall Cabinets can be on the Cabinet Layer, but the lines describing them can be Plotted with different Lineweights if they are assigned appropriate Colors.

**Modifying (changing) Properties (Layer, Color, Linetype):**

As part of this exercise you will continue to allocate certain drawing entities to specific Layers. As part of your Layer management, pay attention to the **Layer** and **Properties Toolbars,** especially to the Current Layer, Color control (ByLayer), and Linetype control (ByLayer) windows. Remember that you can reassign an entity's Layer, Color, and/or Linetype using these toolbars—or use the Properties dialog box.

When building entities via the **Offset, Copy, Mirror** command, etc., those entities will retain the same properties as the original source. So, for example, if you build Interior Walls or Cabinets by **Offsetting** Exterior Walls, you will need to reassign those entities to their proper Layer. Furthermore, you may want to have different Colors and Linetypes on the same Layer. Again, for example, on the Cabinet Layer you may wish to show the Base Cabinets as a Continuous line with a "medium width" color, and the Wall Cabinets as a Hidden line with a "thinner" color.

**Any or all of these modifications (changes) to Properties may be made using the appropriate Toolbar flyout or the Properties dialog box. The simplest way for modifying Layer, Color, and Linetype is to use the Toolbar options.**

To **Modify** (change) an entity's **Layer, select the entity(ies) first,** then select the target Layer using the Toolbar's flyout arrow. To conform to a Layer's "default" settings, let the Color and Linetype Toolbar options read **ByLayer.** The Color and Linetype will automatically adjust to the Layer's settings.

To **Modify Color** and/or **Linetype, select the entity(ies) first,** and then using the Toolbar flyouts, select the desired Color and/or Linetype. You can complete Modifications to any one or all three of these options in one operation.

---

## MODIFYING LAYER, COLOR, LINETYPE WITH TOOLBARS

When modifying properties using the Toolbars, be sure to *select the entity—or entities—FIRST*, then move to the appropriate flyout(s) to make the change. Typically while drawing, you will want the **Color** and **Linetype** controls to read **ByLayer**—or else you will be overriding your default settings.

When you select an entity, its current status will be displayed in the Toolbars—its Layer, Color, and Linetype. If you select a group of entities with different Properties, then the Toolbar displays will appear blank

---

**Modifying Properties with Match Properties Command:**
The **Match Properties** (paintbrush) is a "sometimes" option for changing Properties—but be careful when Matching Properties—remember that not only will Color match, but the Linetype *and Layer* will match as well (unless you use the [S]ettings option at the Command Prompt line).

## DRAW THE INTERIOR FEATURES

**Open** and/or continue working on your ***GRG drawing—your settings should already be locked in. Make sure that you're working in Model Space (check tab at the bottom of the drawing area). Remember that, with what we know to this point, you must go through your Setups for each New drawing, but also that those settings will typically travel with an existing drawing file. If others are sharing your workstation, depending on how the defaults at your station are configured, you may need to verify **OSnap** (and other) settings at the outset of each new workday.

1. Make **WALLS_INTERIOR** the Current Layer, using the Toolbar option.

   Pick the Layer flyout arrow and pick the Layer Name—it will "fly" to the top of the list and display as your CURRENT Layer. This means that any NEW entity you create, will automatically land on the Interior Walls (Partitions) Layer and, by default, will be Yellow, Continuous lines. Remember that lines you Offset, Copy, Extend, etc., will remain true to the properties of their origin—same Layer, Color, Linetype—despite the settings of the CURRENT Layer. You will eventually need to change the properties of such entities.

   **OSnaps** ON, **Ortho** ON

   **DRAW three lines** from the designated inside corners perpendicular (OSnap) to corresponding existing interior walls. These will help align and locate the Guest Bath/Closet and the Dining Room spaces.

2. **Offset** inside of back wall of Bath and Closet to a clear dimension of 5'-0".

   **Offset** interior wall widths 4$^1$/$_2$" at four locations shown.

3. **Offset** Bath/Closet common wall for a clear Bath dimension of 5'-0".

   **Offset** interior wall width 4$^1$/$_2$".

   **Zoom** as required to TRIM to clean up wall intersections.

   **Modify Properties** of interior walls (partitions)—at BATH/CLOS and DINING

   No command—select 4$^1$/$_2$" walls (pick or grab in a crossing window).

   Select WALLS_INTERIOR Layer from Layer Toolbar.

   Color (yellow) and Linetype should default ByLayer.

   Hit the **Escape** key a couple of times to clear screen.

   Though you have two 8"-wide "Interior Walls" running North/South through your building, since these walls would in all probability be masonry—like the Exterior Walls—we are going to leave them on the Exterior Wall Layer, White.

4. **Offset** Closet shelves 12" from each wall. These lines will appear Yellow,

   following the Color and Layer (WALLS_ INTERIOR) of their source.

5. **Trim** the corner—or use **Grips** to clean things up.

   **Modify Properties:** assign shelves to the CABINET Layer.

   Let the COLOR stand as BYLAYER, making them, in our case, BLUE.

6. **Offset** the Centerline of the Water Closet (toilet) 15" from the adjacent partition.

   **Offset** the Centerline of clothes rod in the Closet 2" from the front of the shelf.

   **Modify Properties:** using the Toolbar options, make the following changes:

   Centerline rod: **Layer**—CABINET; **Color**—GREEN; **Linetype**—CENTER

   Centerline W.C.: **Layer**—PLUMBING; **Color**—GREEN; **Linetype**—CENTER

7. **Offset** the Bathroom Vanity (counter/base cabinet) 24"deep × 30"long.

8. **Trim** (or **Fillet**) vanity corner.

   **Extend** as required to meet wall—or use **Grips** with Perpendicular OSnap.

9. **Lavatory (sink): Ellipse (Draw Pulldown or Toolbar)**

   Experiment with the Ellipse command to place an "oval" lavatory in the vanity. Estimate size. Try "massaging" ellipse shape/location using **Grips**—really cool!

   **Water Closet (toilet):**

   **Offset** rear wall and centerline to build 18" × 9" W.C. tank. **Fillet** as required.

   **Offset** rear tank 28" to locate front of bowl; Use **Ellipse/Trim** to complete.

   **Modify Properties:**

   Reassign the Vanity Cabinet to the Cabinet Layer; Color ByLayer.

   Reassign Lavatory and W.C. to the Plumbing Layer; Color ByLayer.

---

**BLOCKS**

In later lessons you will be introduced to shortcut methods for "Inserting Blocks" for items such as Lavatories, Water Closets, Appliances, Doors, and Windows.

Since it is our purpose here to explore and practice specific commands and operations, make sure that you draw these items, including the doors, individually, rather than finding creative ways to increase your efficiency. We'll become more "efficient" down the line.

---

### O.K., LOOKIN' GOOD! NOW FOR THE DOORS.

**Note:** Standard door widths follow a 2" module. The sides of the frame are called *jambs;* the top of the frame is called the *head;* the bottom is called the *sill.*

10. Locate Jambs ("sides" of openings):

    **Offset** walls to locate jambs. **Extend/Trim** (or use **Grips**) to complete.

    **Match Properties** (paintbrush) as needed to match Properties of Jambs to the Properties of their respective wall.

11. Doors: Make the **DOORS Layer Current.**

    **Ortho** ON, **OSnaps** ON—**Draw Doors** (only) to correct length.

    Make sure that you draw the door so that the entire length is "in" the room into which it swings. **DO NOT draw the door over the top of the jamb.**

12. Door Swings:

    Note: The default Arc icon in the Draw toolbar constructs an **Arc** through any three points—this will not produce an accurate arc—you must use an **Arc** option that allows you to select points more accurately.

In the **Draw** Pulldown, the **Arc** flyout offers you a variety of satisfactory options including **Start, Center, End (SCE).** The Center of the arc is the center *point*—like the center point of a circle—where you would place the metal point of a compass to construct the arc or circle. It is not a point "on" the arc. It will *always* fall at the "jamb end" of the door...remember that.

AutoCAD typically defaults to a *counterclockwise* direction when drawing arcs. That means the **Start** point and **End** point will vary, according to the orientation of the specific door.

You may use the "Arc through three points" icon for constructing your door swings by forcing the command to locate to the **Center, Start,** and **End** points. Select the standard icon from the **Draw Toolbar** and then **type C** (for Center) and **Enter.** Following the prompts, build the arc by selecting the base of the door (at jamb) as **Center,** then the **Start** and **End** points of the arc. Remember that the arc will build *counterclockwise* from whichever point you select as the starting point.

**So...Draw Arcs...OSnaps ON**

**Select Arc (SCE)** from Draw Pulldown and follow prompts.

...**OR**...

Select **Arc(3pts)** icon from Draw Toolbar; type "C" and Enter; follow prompts.

**Modify Properties:**

We want for the swings to remain on the DOORS Layer, but want them to plot "thinner" than the Doors themselves. So we will assign them a "thin" color—RED. Remember that we have assigned different line widths to different colors—Yellow (the Doors themselves) will be Medium Wide and Red (the Swings) will be the thinnest.

13. Kitchen Cabinets and Buffet: **Offset, Fillet, Trim, Grips**...

24" deep Base Cabinets, 12" deep Wall Cabinets

Locate Island by 4' aisle and Buffet alignment

**Modify Properties:**

Assign all Base Cabinets/Island to CABINET LAYER; Color ByLayer

Wall Cabinets to CABINET Layer; Color GREEN; Linetype HIDDEN

Are you starting to get the color thing? We'll tell AutoCAD to plot Blue lines (Base Cabinets) wider than Green lines (hidden lines for Upper Cabinets).

## YOUR DRAWING IS STARTING TO LOOK REALLY DOWNTOWN.
## GIVE YOURSELF A BIG PAT ON THE BACK.

### Save your drawing. And of course you've been saving periodically as you drew—right?

Now the Plot thickens...literally! We'll set Lineweights (Plot Style) so you can better grasp the idea of Colors vs. Lineweights. Most of the setup will be the same as before...we're just going to add a step or two. Refer to *Supplement 6: Quick Guide for Plotting in Paper Space* as you go through this process...start to become familiar with this guide.

## PLOT THE FLOOR PLAN: INTERIOR FEATURES IN PAPER SPACE
Use the **Quick Guide** as reference.

The following setup will produce a "check plot"—a small scale plot for quick review. Your instructor may want you to produce a larger plot. Verify the procedure with your instructor.

In **Format** Pulldown, **select Linetype.**

"**Deselect**" box "**Use paper space units for scaling**" if you have not done so.

This box must be turned off for Center and Hidden lines to show up in Paper Space.

**Select the Layout Tab** you've previously configured to enter Paper Space.

| | |
|---|---|
| **Page Setup:** | Right click on the Layout Tab for Page Setup. |
| **Plot Device Tab:** | (at top of dialog box) |
| **Plotter Configuration:** | per your instructor |
| **Plot Style Table:** | **monochrome**—select **Edit** set **Lineweights** for colors 1–7 per **Quick Guide** for SMALL PLOT. Watch decimal points! Color 7, Black, is the same as White. Don't worry about Grays at the this time. Select **Save and Close** at the bottom of the dialog box. |
| **Layout Settings Tab:** | (at top of dialog box) |
| **Paper Size:** | letter (8.5 × 11) |
| **Orientation:** | landscape |

Other settings are probably OK…your instructor will verify.

**OK** Page Setup and you'll see your designated piece of paper.

Create a **single Viewport** nearly the size of your paper boundaries (Viewport may already be there).

**Activate** and **Scale** the Viewport to ⅛" = 1'-0"; center the drawing in the Viewport (**Pan**).

(Plans are typically plotted a ¼" = 1'-0", "check plots" are often run at smaller scales).

Place the Viewport on the **VIEWPORT LAYER** (set for "no plot").

**Save** your finished setup.

**Right click** on the Layout Tab to **Preview** and **Plot.**

**Save** to the drive location designated by your instructor and create a Backup file on the (A:) Drive per your instructor's direction.

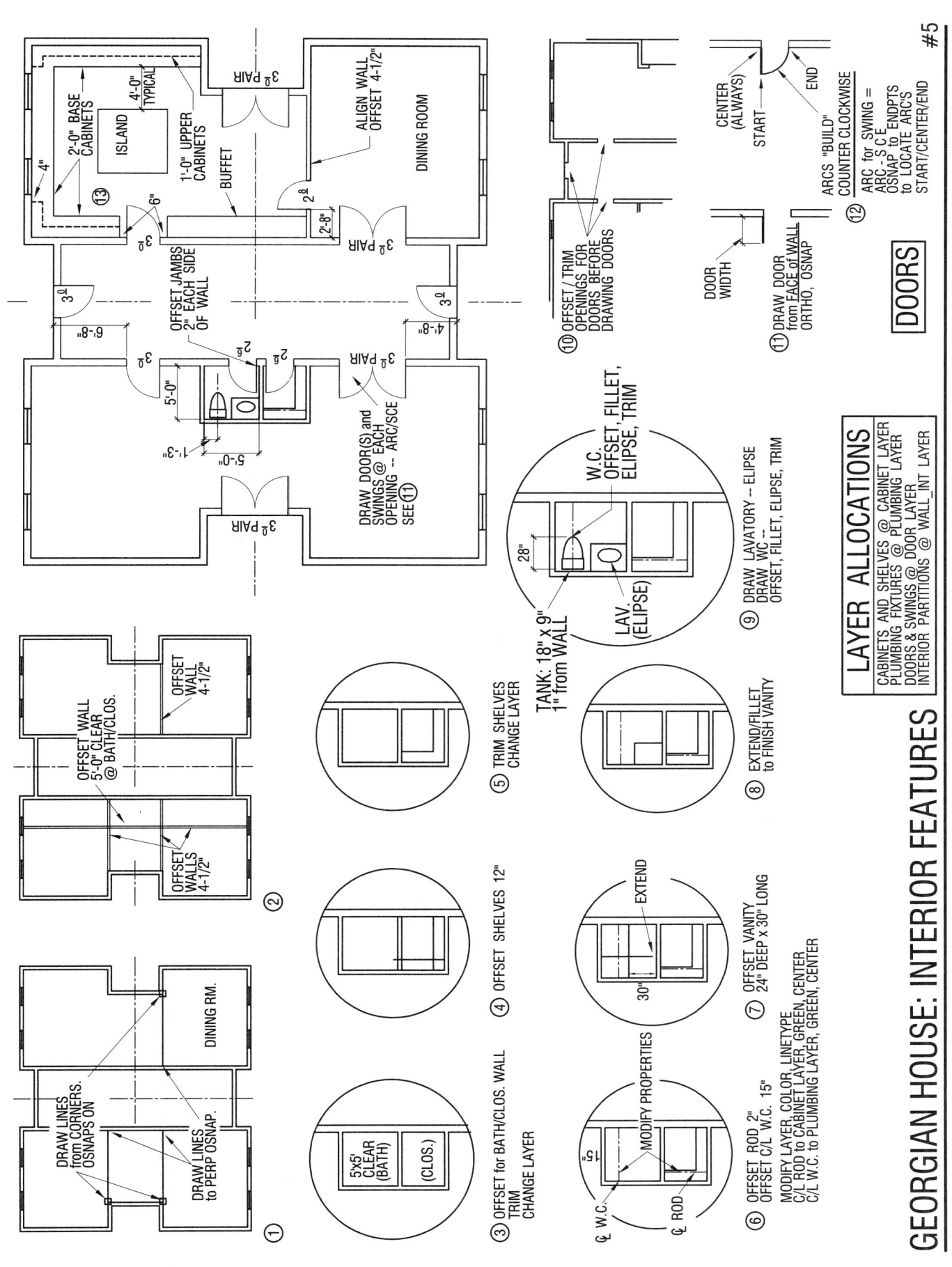

# GEORGIAN HOUSE: INTERIOR FEATURES

#5

## LAYER ALLOCATIONS

CABINETS AND SHELVES @ CABINET LAYER
PLUMBING FIXTURES @ PLUMBING LAYER
DOORS & SWINGS @ DOOR LAYER
INTERIOR PARTITIONS @ WALL_INT LAYER

### DOORS

⑩ OFFSET / TRIM OPENINGS FOR DOORS BEFORE DRAWING DOORS

⑪ DRAW DOOR from FACE of WALL ORTHO, OSNAP
DOOR WIDTH

⑫ ARCS "BUILD" COUNTER CLOCKWISE
CENTER (ALWAYS)  START  END
ARC for SWING = ARC - S C E
OSNAP to ENDPTS to LOCATE ARC'S START/CENTER/END

---

① DRAW LINES from CORNERS. OSNAPS ON
DRAW LINES to PERP OSNAP.
DINING RM.

② OFFSET WALL 5'-0" CLEAR @ BATH/CLOS.
OFFSET WALL 4-1/2"
OFFSET WALLS 4-1/2"

③ OFFSET for BATH/CLOS. WALL TRIM CHANGE LAYER
5x5' CLEAR (BATH)  (CLOS.)

④ OFFSET SHELVES 12"

⑤ TRIM SHELVES CHANGE LAYER

⑥ OFFSET ROD 2"
OFFSET C/L W.C. 15"
MODIFY LAYER, COLOR, LINETYPE
C/L ROD to CABINET LAYER, GREEN, CENTER
C/L W.C. to PLUMBING LAYER, GREEN, CENTER
℄ W.C.  ℄ ROD  15"  MODIFY PROPERTIES

⑦ OFFSET VANITY 24" DEEP x 30" LONG
30"

⑧ EXTEND/FILLET to FINISH VANITY
EXTEND

⑨ DRAW LAVATORY -- ELIPSE
DRAW WC --
OFFSET, FILLET, ELIPSE, TRIM
W.C. OFFSET, FILLET, ELIPSE, TRIM
TANK: 18" x 9" 1" from WALL
28"
LAV. (ELIPSE)

---

⑬ 2'-0" BASE CABINETS
4" 
ISLAND  4'-0" TYPICAL
1'-0" UPPER CABINETS
BUFFET
3⁰ PAIR
ALIGN WALL OFFSET 4-1/2"
DINING ROOM
6"  3⁰
OFFSET JAMBS 2" EACH SIDE OF WALL
2⁶  2⁶  2-8"  2⁸  3⁰ PAIR
3⁰  6'-8"  3⁰
5'-0"  1'-3"  5'-0"
DRAW DOOR(S) and SWINGS @ EACH OPENING -- ARC/SCE SEE ⑪
3⁰ PAIR  4'-8"  3⁰
3⁰ PAIR

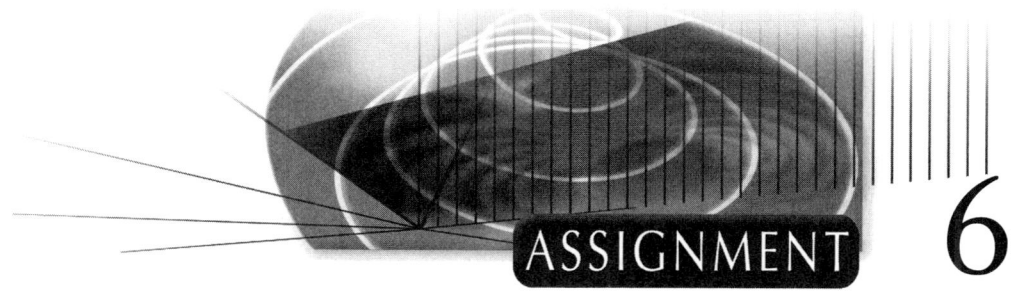

# Shutter

## COMPETENCIES/LEARNING OUTCOMES

Upon successful completion of the **Shutter,** the student will have used and begun to master the following settings, operations, and drawing commands.

**Settings:** Units, Limits, Snap, Grid, Polar Tracking, Ortho, OSnap, Layers, Linetypes

**Operations:** Open/Exit AutoCAD; Access Menus; Enter Commands; Startup Wizard; Open; Save As and Save drawing files; Creating/Saving backup files.

**Commands:**

Line

Offset

Fillet (0)

Erase: Select, Window, Crossing Window

Undo

Zoom Window, Previous, All, Realtime

Pan

Save

Save As

Plot in Paper Space

Trim

OSnap

Fillet with Radius (R) setting

Grips

Circle

Copy and Multiple Copy

Mirror

Layer Properties

Linetype Load/Selection

Linetype Scale

Extend

Properties (Modify and Match)

Arc: SCE (Start, Center, End)

**(new) Array (Rectangular)**

## PROCEDURE

Working alone and with classmates, incorporate information from lectures, demonstrations, and corresponding information in your text to complete this drawing. In this, your sixth assignment, you will continue to reinforce lessons/commands encountered in your previous work. If you are using a supplemental text(s), general reviews will serve you well. Now that you've begun to understand how AutoCAD "thinks," these reviews will cement many items into place and expand your insight as well.

In this and remaining assignments, you will be receiving less and less step-by-step instruction. In turn, you will rely more and more on your own developing skills and resources. Remember to draw on your previous lessons to solve new problems. Start to explore Toolbars and Pulldowns to discover new options. Ask your instructor and fellow classmates for help if you're stuck, but try to solve problems on your own first—just as you would in an office situation.

As you continue to build your expertise, think of different strategies or approaches one might use to create a specific drawing. Try to find the most efficient way for you. As always …

### RELAX, EXPLORE, ASK QUESTIONS,
### WORK WITH YOUR FELLOW CLASSMATES.

### SHUTTER

You will begin a New drawing for the SHUTTER, so on a "clean screen" from the **Create New Drawing** dialog box …

Pick    **Use a Wizard**

**Quick Setup**

**Units: Architectural** … **Next**

**Area:** Width = 11', Length = 8'6—make sure you include the "foot mark"

**Finish**

**Zoom All** to ensure that you've captured your entire screen

Pick **Save As** (from File Pulldown)

**Save In:** Locate your designated drive—verify with your instructor.

**File Name:** Your three initialsSHUTTER—ex: mrbSHUTTER

>    Remember to **SAVE periodically while working** and **SAVE before you Exit** AutoCAD

Other Settings:
Tools/Drafting Settings: **Snap and Grid** tab: Grid 4"

>    **OSnap tab:**  Verify OSnap Settings—Endpoint, Midpoint, Center, Intersection, Perpendicular ON Other options OFF.

>    **Always verify these settings with each new drawing!**

Verify **Fillet Radius** = 0" Always verify Fillet Radius with each new drawing!

**Toggles: Snap** OFF; **Grid** ON; **Ortho** ON; **OSnap** ON

### DRAW the SHUTTER!

1. Draw the outline of the shutter 1'-8" wide × 4'-0" high.

2. **Offset** 2" frame to "inside"; **Trim** or **Fillet** 0" corners.

   **Modify Properties:** Grab the entire frame in a Window and change its Color to Magenta. Make this change using the Toolbar—not the dialog box.

   Leave the entities on **Layer 0.**

## ARRAY RECTANGULAR AND ARRAY POLAR

The **Array** command will make "mass copies" on either rectangular grids **(Rectangular Array)** or around a center point **(Polar Array)**. Draw a single chair and you can create an entire theater with columns and rows of chairs spaced a specific distance apart, or you can space any number of chairs evenly around a circular table.

For the Shutter we need to create 2" louvers. We could **Offset** each slat 2" from the next, we could do a **Copy** or **Multiple Copy,** or we can do an **Array**—not a spectacular "stadium-filling" **Array,** but just a single column with the desired number of rows. That's one stack (the column) of 21 louvers (the rows).

3. **Array** the louvers. From the **Modify** Pulldown or Toolbar ...

Select **Array** ... the **Array** dialog box will appear.

Select **Rectangular.**

Pick **Select Objects** icon.

On your drawing screen, select the **bottom inside frame.**

**Right click** or **Enter** to tell AutoCAD you're done selecting objects.

**Rows: 22** ... that's the number of Louvers—but notice that we needed to include the object we selected to array as part of this count.

**Columns: 1** ... we just want a single stack here.

**Row Offset: 2** ... We want the rows (the louvers) to be 2" apart.

**Preview** to see the results of your entries.

**Accept** if you're happy. (**Modify** will allow you to make changes.)

**Modify Properties:** Make the 21 louvers RED (Layer 0).

So now you should have a 2" Magenta frame and 21 Red louvers.

4. **OSnap** a layout line, **Mid**point side to **Mid**point side.

We'll use this line to locate the center points of the circles at the center panel.

Remember, when drawing on the computer, to give yourself the same types of layout lines that you need "on the boards." You can erase them later. Some drafters even create a Layer called LAYOUT on which they put and "leave" important layout lines. That Layer is then turned OFF when the drawing is plotted. In this instance, you need to draw circles with center points at the "middle" of the shutter panel—you could "count louvers" to find the center—or quickly OSnap a layout line side to side as the example shows.

5. **Circle** command, using the Midpoint OSnap for center point locations ...

Type and Enter Radius values. Complete this process once to create an 8" DIA circle and again for a 4" DIA circle. Don't miscue and enter Diameters when AutoCAD is asking for Radii.

6. **Erase** the layout line and the "center" louver.

7. **Trim** out the center panel interior, select the outer circle as your Cutting Edge.

8. **Match Properties** (paint brush) ... change center panel lines to Magenta to match the frame.

Your shutter's frame and center panel should now be MAGENTA and the louvers should be RED. Remember that we can designate the Plot's line thicknesses by the Color of the line. We will be plotting MAGENTA as a "medium" thickness line, and RED as a "thin" line. This means that your Plot will "read" with two distinct Lineweights.

### NEAT—HUH?

**Save** your drawing to your designated location.

## PLOT THE SHUTTER IN PAPER SPACE

Use the **Quick Guide** as needed.

**Select a Layout Tab** to enter Paper Space:

| | |
|---|---|
| **Page Setup:** | Right click on the Layout Tab for Page Setup. |
| **Plot Device Tab:** | (at top of dialog box) |
| **Plotter Configuration:** | per your instructor. |
| **Plot Style Table:** | monochrome—select Edit—set Lineweights for colors 1 Red and 6 Magenta per **Quick Guide** for SMALL PLOT. Watch decimal points! Select **Save and Close** at the bottom of the dialog box. |
| **Layout Settings Tab:** | (at top of dialog box) |
| **Paper Size:** | letter (8.5 × 11) |
| **Orientation:** | landscape |

Other settings are probably OK...your instructor will verify.

OK Page Setup and you'll see your designated piece of paper.

Create a **single Viewport** nearly the size of your paper boundaries.

**Scale** the Viewport to 1" = 1'-0"; center the Block Out in the Viewport (**Pan**).

Place the Viewport on the **VIEWPORT LAYER** (make a new layer: "Viewport"—no plot)

**Save** your finished setup.

**Right click** on the Layout Tab to **Preview** and **Plot.**

**Save** to the drive location designated by your instructor and create a Backup file on the (A:) Drive per your instructor's direction.

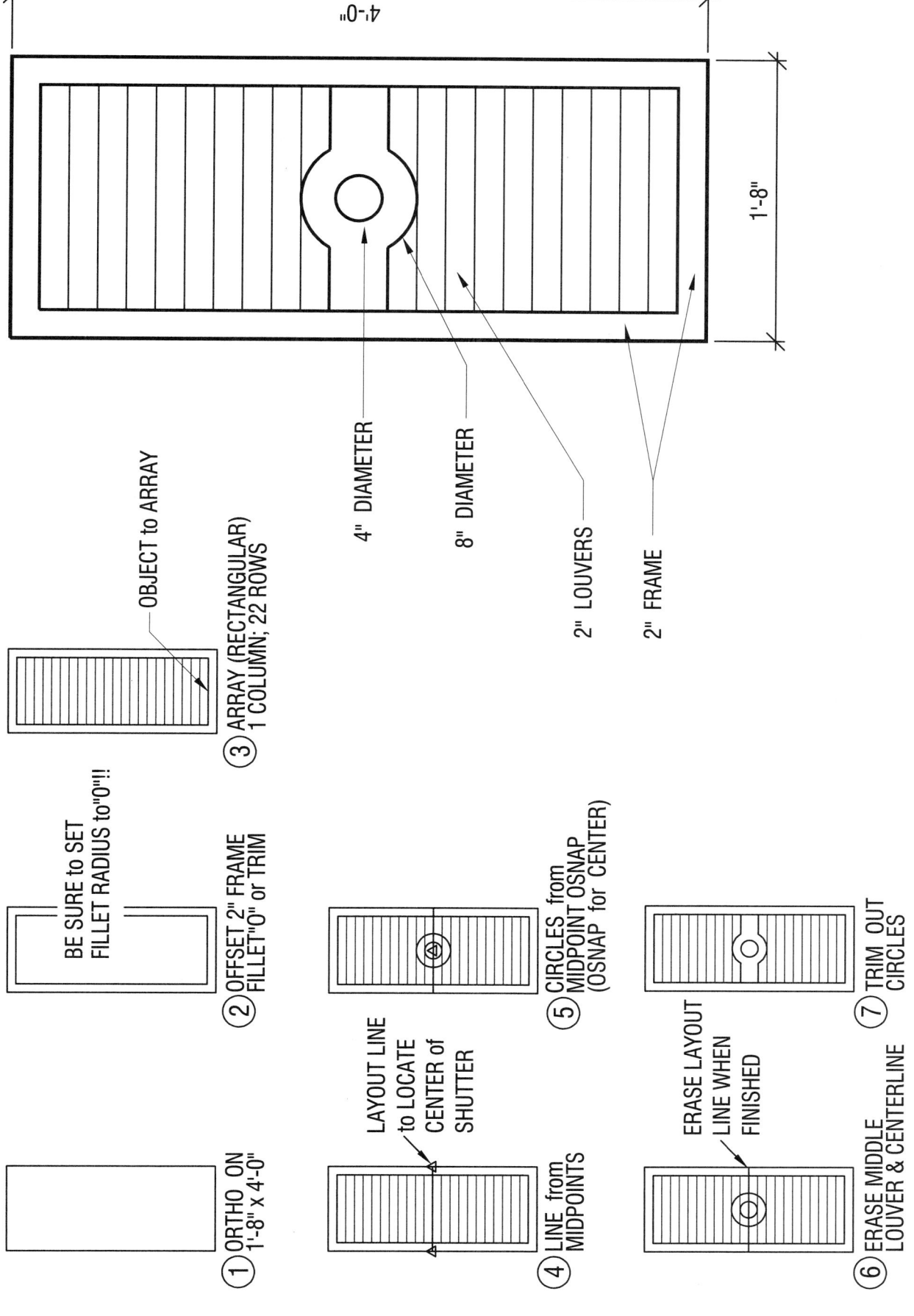

4'-0"

1'-8"

4" DIAMETER

8" DIAMETER

2" LOUVERS

2" FRAME

OBJECT to ARRAY

③ ARRAY (RECTANGULAR)
1 COLUMN; 22 ROWS

BE SURE to SET
FILLET RADIUS to"0"!!

② OFFSET 2" FRAME
FILLET"0" or TRIM

LAYOUT LINE
to LOCATE
CENTER of
SHUTTER

⑤ CIRCLES from
MIDPOINT OSNAP
(OSNAP for CENTER)

ERASE LAYOUT
LINE WHEN
FINISHED

⑦ TRIM OUT
CIRCLES

① ORTHO ON
1'-8" x 4'-0"

④ LINE from
MIDPOINTS

⑥ ERASE MIDDLE
LOUVER & CENTERLINE

SHUTTER

#6

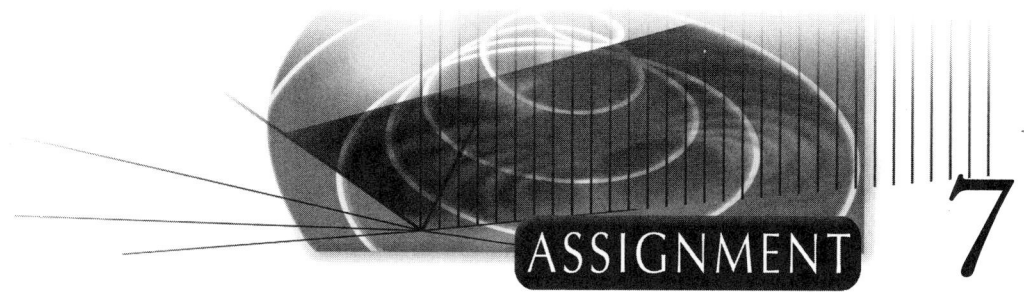

# Staircase

## COMPETENCIES/LEARNING OUTCOMES

Upon successful completion of the **Staircase,** the student will have used and begun to master the following settings, operations, and drawing commands.

**Settings:** Units, Limits, Snap, Grid, Polar Tracking, Ortho, OSnap, Layers, Linetypes

**Operations:** Open/Exit AutoCAD; Access Menus; Enter Commands; Startup Wizard; Open; Save As and Save drawing files; Creating/Saving backup files.

**Commands:**

Line

Offset

Fillet (0)

Erase: Select, Window, Crossing Window

Undo

Zoom Window, Previous, All, Realtime

Pan

Save

Save As

Plot in Paper Space

Trim

OSnap

Fillet with Radius (R) setting

Grips

Circle

Copy and Multiple Copy

Mirror

Layer Properties

Linetype Load/Selection

Linetype Scale

Extend

Properties (Modify and Match)

Arc: SCE (Start, Center, End)

Array (Rectangular)

**(new)** **Break at Point**

## PROCEDURE

Working alone and with classmates, incorporate information from lectures, demonstrations, and corresponding information in your text to complete this drawing. As always …

### RELAX, EXPLORE, ASK QUESTIONS,
### WORK WITH YOUR FELLOW CLASSMATES.

## STAIRCASE

A New drawing for the **STAIRCASE**—from the **Create New Drawing** dialog box …

Pick     **Use a Wizard**

Pick     **Quick Setup**

**Units: Architectural** … **Next**

**Area:** Width = 29'4, Length = 22'8—make sure you include the "foot mark"

**Finish**

**Zoom All** to ensure that you've captured your entire screen

Pick **Save As** (from File Pulldown)

**Save In:** Locate your designated drive—verify with your instructor.

**File Name:**   Your three initialsSTAIRCASE—ex: mrbSTAIRCASE

   Remember to **SAVE periodically while working** and **SAVE before you Exit** AutoCAD.

**Other Settings:**

**Format** Pulldown/**Linetype –Load** Hidden Line; **Global Scale Factor** = 24;

   **Deselect** "Use paper space units for scaling."

**Tools/Drafting Settings: Snap and Grid** tab: Grid 4"—no Snap

   **OSnap** tab: Verify OSnap Settings—End pt., Mid pt., Center, Intersection, Perpendicular ON.
   Other options OFF.

   **Always verify these settings with each new drawing!**

Verify **Fillet Radius** = 0" Always verify Fillet Radius with each new drawing!

**Toggles: Snap** OFF; **Grid** ON; **Ortho** ON; **OSnap** ON

Again, we are now using the Grid as a visual reference only, to "see" where our paper is. It is not necessary that the Grid be ON when drawing. Many prefer, once the object is "located" on the paper, to turn the Grid OFF. Try it—you might like having a *clean, mean screen* on which to work.

## DRAW the STAIRCASE!

   1. Ortho ON—Drag and Draw the outline of the staircase 4'-6" wide × 15'-0" long.

   2. **Offset** 2" Hand Rail to outside of stair.

      **Array (Rectangular)** 12" treads

         Pick **Rectangular; Select Objects** (select bottom line);

         **Rows** 15; **Columns** 1; **Row Offset** 12"

   3. Ortho OFF, draw BREAK LINE as shown, a little past the staircase's center.

      "Eyeball" this baby.

The rail and stringers (sides of stairs) are each single, continuous lines. In Architectural drawings, on a ground floor we typically show the lower portion of a flight of stairs with Continuous lines and the remaining, upper portion with Hidden lines (if we show the upper portion at all). This means that we need to change "part" of these Continuous lines to Hidden lines, which we can't do unless we "break" the single line into two separate lines. So here we go ... the **Break** command.

## BREAK COMMAND

First off, don't confuse a "Break Line," a typical Architectural convention, with the **Break** command. These are coincidental terms. A "Break Line" is drawn to indicate that the object we're drawing is broken away—we're only showing a portion of the entire object. The **Break** command is used to break a section or "piece" from an entity (**Break at Two Points**) or to break a single entity into two or more separate pieces (**Break at Point**).

To break a line at a single point—to divide the line into two pieces that remain touching end to end—opt for the **Break @ Point** icon in the **MODIFY Toolbar**. Accessing **Break** from the Modify pull-down requires additional steps to accomplish the same end.

4. **Break** the rail and stringer—three lines; **Break** treads where they cross break line.

    Select **Break @ Point** from **Modify** Toolbar.

    Select **Object to Break**—anywhere on one of the railing or stinger lines.

    > You can only do one line at a time—you'll need to repeat the command.

    > **OSnap** ON—Select Intersection between the selected object and the drawn "Break Line." The single line is now two lines, end to end.

    > To check—Just pick the line with no active command. The Grips will tell of your success.

    > Repeat this process for the remaining two vertical lines.

    > Where you **Break** the treads depends on the location of your break line.

    **Modify Properties:**

    > Change ALL lines "above" the Break Line to HIDDEN, GREEN.

    > Change Railing to MAGENTA.

    Change remaining treads and stringer to BLUE.

    > Change the "Break Line" to RED.

    > Let all lines remain on **Layer 0.**

5. Draw the Newel Post (that thing you absolutely, positively don't want to hit when you slide down the banister at grandma's house!!!)

    **Locate Centerlines** for Circles:

    > **Offset** outside rail 2" for vertical layout line.

    > **Ortho** ON, **draw a line** extending from the bottom tread for horizontal layout line.

    > The intersection of these lines will provide the center point for the circles.

    **Circle** command ... this is cool ... **OSnap to center point** formed by layout lines.

    > Instead of typing in the Radius or Diameter values—just **"drag" the circle** to the appropriate OSnap at the rail and "pick" to drop the circle in perfect alignment. Do this twice, once for each circle.

6. **TRIM** out the Newel per your example.

    > **Save** your drawing to your designated location.

## PLOT THE STAIRCASE IN PAPER SPACE

Use the **Quick Guide** as needed.

**Select a Layout Tab** to enter Paper Space:

| | |
|---|---|
| **Page Setup:** | Right click on the Layout Tab for Page Setup. |
| **Plot Device Tab:** | (at top of dialog box) |
| **Plotter Configuration:** | per your instructor |
| **Plot Style Table:** | **monochrome**—select **Edit**—set Lineweights 1 Red, 3 Green, 5 Blue, and 6 Magenta per **Quick Guide** for SMALL PLOT. Watch decimal points! Select **Save and Close** at the bottom of the dialog box. |
| **Layout Settings Tab:** | (at top of dialog box) |
| **Paper Size:** | letter (8.5 × 11) |
| **Orientation:** | **landscape** |

Other settings are probably OK … your instructor will verify.

**OK** Page Setup and you'll see your designated piece of paper.

Create a **single Viewport** nearly the size of your paper boundaries.

      **Scale** the Viewport to $^3/_8$" = 1'-0"; center the Block Out in the Viewport (Pan).

      Place the Viewport on the **VIEWPORT** Layer (make new layer: "Viewport"—no plot)

      **Save** your finished setup.

**Right click** on the Layout Tab and select **Plot.**

      **Save** to the drive location designated by your instructor and create a Backup file
          on the Floppy (A:) Drive per your instructor's direction.

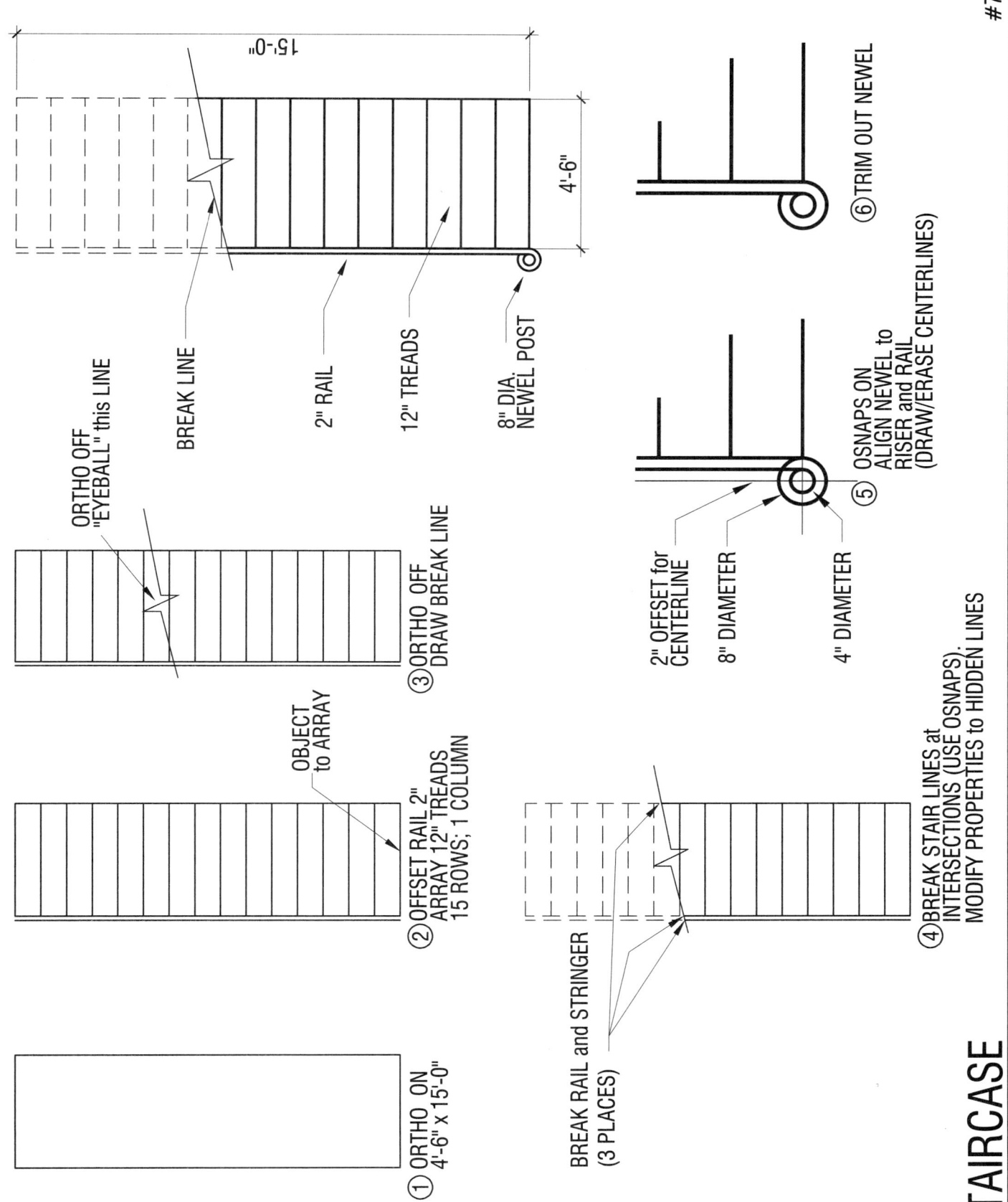

① ORTHO ON
4'-6" x 15'-0"

② OFFSET RAIL 2"
ARRAY 12" TREADS
15 ROWS; 1 COLUMN

OBJECT
to ARRAY

③ ORTHO OFF
DRAW BREAK LINE

ORTHO OFF
"EYEBALL" this LINE

④ BREAK STAIR LINES at
INTERSECTIONS (USE OSNAPS).
MODIFY PROPERTIES to HIDDEN LINES

BREAK RAIL and STRINGER
(3 PLACES)

15'-0"

4'-6"

BREAK LINE

2" RAIL

12" TREADS

8" DIA.
NEWEL POST

⑤ OSNAPS ON
ALIGN NEWEL to
RISER and RAIL
(DRAW/ERASE CENTERLINES)

⑥ TRIM OUT NEWEL

2" OFFSET for
CENTERLINE

8" DIAMETER

4" DIAMETER

STAIRCASE

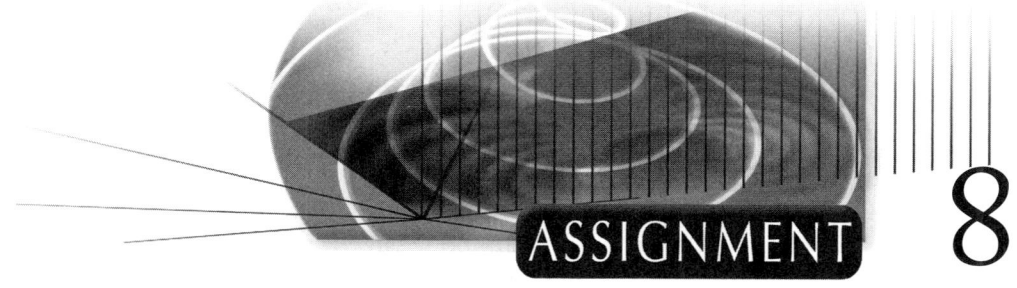

# Fireplace, Furniture, Appliances

## COMPETENCIES/LEARNING OUTCOMES

Upon successful completion of the **Fireplace, Furniture,** and **Appliances,** the student will have used and begun to master the following settings, operations, and drawing commands.

**Settings:** Units, Limits, Snap, Grid, Polar Tracking, Ortho, OSnap, Layers, Linetypes

**Operations:** Open/Exit AutoCAD; Access Menus; Enter Commands; Startup Wizard; Open; Save As and Save drawing files; Creating/Saving backup files.

**Commands:**

Line

Offset

Fillet (0)

Erase: Select, Window, Crossing Window

Undo

Zoom Window, Previous, All, Realtime

Pan

Save

Save As

Plot in Paper Space

Trim

OSnap

Fillet with Radius (R) setting

Grips

Circle

Copy and Multiple Copy

Mirror

Layer Properties

Linetype Load/Selection

Linetype Scale

Extend

Properties (Modify and Match)

Arc: SCE (Start, Center, End)

Array (Rectangular)

Break at Point

**(new)   Rotate**

**Array (Polar)**

## PROCEDURE

Working alone and with classmates, incorporate information from lectures, demonstrations, and corresponding information in your text to complete this drawing.

You will be pretty much on your own for this stuff—a good chance to implement your former lessons. However, remember to ask questions if you find yourself stumped. And, as always…

RELAX, EXPLORE, ASK QUESTIONS,<br>WORK WITH YOUR FELLOW CLASSMATES.

## FIREPLACE, FURNITURE, APPLIANCES

**Open** and/or continue working on your \*\*\*GRG drawing (FLOOR PLAN)—your settings should already be locked in…

**Snap** OFF; **Ortho** ON; **OSnap** ON; **Fillet** Radius = 0".

**Complete these drawings "off to the left" of your Floor Plan, Zooming in on workable views. Don't worry about staying within your Drawing Limits (Grid). In the next assignment, you will move the items into your floor plan.**

**Save** your work periodically—this should be second nature to you by now!!!

### 1. FIREPLACE

Draw the Fireplace on the **WALL_EXTERIOR** Layer (make it "Current").

**Ortho** ON, Draw the 6'-0" × 2'-8" outline.

**Offset/Fillet/Trim** to build firebox.

For the angled (splayed) firebox walls, **Offset** vertical and horizontal lines to locate endpoints, then OSnap lines to those intersections for the angled lines—like "connecting the dots."

**Fillet/Trim/Grips** to make corners, **Erase** layout lines. There's a Fireplace in there somewhere!

Remember that the **Fillet** will extend and/or trim angled lines to make corners.

**Modify Properties:**

4" firebrick lining—BLUE

Line (joint) @ front of firebox opening—BLUE

You'll either have to **Trim** and draw a new line or **Break@Point** the line at the front of the firebox to change it to BLUE.

Set the **FURNITURE** Layer as "Current" Layer. Draw the Furniture as shown.

### 2. SOFA; COFFEE, SIDE, and SOFA TABLES

**SOFA:** To **Fillet** the corners (set 1.5" Radius), you will either have to draw each cushion "individually"—"clicking off" separate line segments as you draw, OR if you create the SOFA with continuous lines, you will have to **Break @ Point** at INTERSECTIONS (OSnap) to successfully **Fillet** rounded corners.

**TABLES: Reset Fillet Radius to 0"** for squared corners. Draw one SIDE TABLE at the desired location adjacent to SOFA and **Mirror** the second, using Midpoint-to-Midpoint OSnaps on SOFA for Mirror Line.

3. **HALL TABLE and SIDE CHAIRS**

   **SIDE CHAIRS: Offset** the 16" Circle 2" for the CHAIR back.

   Draw one angled line ("eyeball" with Ortho OFF) from the CENTER OSnap "beyond" the outer circle. **Mirror** that line, using CENTER OSnap (Ortho ON) as the start point for the Mirror Line.

   **Trim** to complete. **Mirror** or **Copy** second CHAIR.

4. **HUTCH**

   Draw outline, then construct details with the **Offset, Trim/Fillet.**

5. **CHAIR**

   Follow the illustrations. You will eventually **Copy** this CHAIR several times.

6. **DINING TABLE and CHAIRS**

   Draw the DINING TABLE 8'-0" × 3'-6".

   **Offset** layout lines to mark CHAIR positions.

   Using **Grips** (drag and type), extend one vertical layout line 12" beyond table.

## DRAGGING AND TYPING WITH GRIPS

Note that you can drag a Grip in the desired direction and "type/Enter" the distance to extend or shorten—similar to dragging and typing when drawing lines.

**Copy** the CHAIR, OSnap Midpoint of its back as **Base point**; Endpoint of vertical layout line at table as **Displacement point.**

**Trim/Erase** portions of CHAIR under the TABLE.

You may **Zoom** IN and OUT as required "during" the **Copy** command for accuracy. In other words, you may Zoom IN on the CHAIR to "grab the Base Point," then **Zoom** OUT and Zoom back IN on the TABLE to position the CHAIR.

**Copy** the CHAIR to the same side of the TABLE, using Endpoint to Endpoint

OSnaps (layout lines @ Table) as **Base** and **Displacement points.**

**Mirror** the two CHAIRS to the opposite side of the TABLE.

Use horizontal layout line as guide for Mirror Line, OSnapping to its ends.

**Copy** one CHAIR "off to the side" to place "end CHAIR"

**Rotate** (a NEW command) the CHAIR to the proper orientation.

   Follow command prompts on your screen:

   **Base Point**—select the Midpoint of the CHAIR back.

   **Rotation Angle**—type and Enter 90 (to Rotate 90° counterclockwise).

## ROTATE COMMAND

Think of the Base Point as a "pin" around which the object will spin—as if you tacked the object to the wall and gave it a twirl. If you Rotate with Ortho ON, the computer will automatically "lock" you into one of the four 90° rotation options as you manipulate your mouse and left click to reposition the object.

For "odd" angles of rotation—or to ensure accuracy—you can Type and Enter the exact angle you wish. Remembering that the computer measures angles counterclockwise, typing and Entering 45 will Rotate the object 45° to the left, typing and entering–45 (minus 45) will Rotate the object 45° to the right.

Using **Grips** (drag and type), extend the horizontal layout line 12" beyond table.

**Move** the Rotated CHAIR into place—OSnap Midpoint Chair back (Base Point) to Endpoint layout line (Displacement Point).

**Mirror** Chair to opposite end of table.

Mirror Line = Midpoint-to-Midpoint OSnap across width of TABLE.

**Erase** layout Lines.

**You're saving your work as you go—right!?**

7. **BREAKFAST TABLE and CHAIRS**

Draw 42" Diameter TABLE—if AutoCAD wants a Radius—you must convert (or type **D** and **Enter** to use Diameter).

Draw a VERTICAL layout line (Ortho ON) from Center OSnap 33" long.

This will position the endpoint of the line 12" beyond the circle—Radius = 21", right?

**Copy** a CHAIR into position using OSnaps as Base and Displacement Points.

**Trim/Erase** portions "below" TABLE; including layout line.

**And now for one of the NEATEST of all the commands—Polar Array**

**Array** command—select and **Array** dialog box will appear.

Select **Polar.**

Pick **Select Objects** button—grab CHAIR in a window—**Enter**

**Center Point**—at *x/y* coordinates line, select "**Pick Center Point**" button.

  Pick CENTER of circular TABLE—OSnap to center for accuracy!

**Total Number of Items** = 5 (you must count the "original" as part of the Total Number).

**Angle to Fill** = 360

**Rotate Items as Copy**—select/check the box at lower left of the dialog box.

**Preview**—**Accept** if things look good—**Modify** if you need to adjust.

---

### MORE ON THE ARRAY COMMAND

OKAY—the **Polar Array** copies items "around a center point"—the **Rectangular Array** copies items on a rectangular grid. In both cases you must count the original item as one of the total. In the example you just completed, you could have placed chairs around "half" the table by selecting an Angle to Fill of 180° (a half circle), rather than 360° (a full circle). Remember in that case, entering 180 would have filled a rotation counterclockwise in direction—filled the left side of the table. Entering a value of –180 (minus 180) would have filled the right-hand side of the table.

Typically in **Polar Arrays,** we want to Rotate Items as they copy—the box in the lower left corner of the dialog box. If you had not rotated the CHAIRS as they copied, they would have ended up sideways or backward to the table—right?

---

8. **SINK, RANGE/OVEN (R/O), REFRIGERATOR (REF)**

**Make PLUMBING Layer "Current"**—draw SINK.

  **Offset** compartments 3" at back; 2" at front, sides, & between.

  Change compartment outlines and drains to RED.

**Make APPLIANCE Layer "Current"**—draw R/O (range/oven) and REF (refrigerator).

Small burners = $3^1/_2$" Radius, Large = $4^1/_2$" Radius.

Change burners on R/O to RED.

**Offset** Refrigerator door 3"—change offset line to RED.

## PLOT THE FIREPLACE, FURNITURE, APPLIANCES IN PAPER SPACE

Use the **Quick Guide** as needed.

**Select a Layout Tab** to enter Paper Space:

| | |
|---|---|
| **Page Setup:** | Right click on the Layout Tab for Page Setup. |
| **Plot Device Tab**: | (at top of dialog box) |
| **Plotter Configuration:** | per your instructor |
| **Plot Style Table:** | monochrome—select Edit—set Lineweights 1 Red, 4 Cyan, and 6 Magenta per **Quick Guide** for SMALL PLOT. Watch decimal points! Select **Save and Close** at the bottom of the dialog box. |
| **Layout Settings Tab:** | (at top of dialog box) |
| **Paper Size:** | letter (8.5 × 11) |
| **Orientation:** | landscape |

Other settings are probably OK…your instructor will verify.

**OK** Page Setup and you'll see your designated piece of paper.

All right—this is something new; you're going to plot a part(s) of the entire drawing—the Fireplace/Furniture/Appliances—but not the Floor Plan. And we're not going to plot these to any particular scale—we just want to have a look for a visual "check." Plotting "to fit" the Viewport.

Create a **single Viewport** nearly the size of your paper boundaries.

Activate the Viewport—double click inside or toggle

**Pan** (Mr. Hand) so that your FIREPLACE, FURNITURE, APPLIANCES are "centered" in the Viewport. You may need to do some "arranging" in Model Space (Move command) to group these items more tightly.

**Zoom Realtime,** "filling" the Viewport with your Furniture to the maximum size allowed.

Alternate back and forth between **Pan** and **Zoom Realtime** to maximize the view.

Make sure that your Viewport is on the VIEWPORT Layer (no Plot).

**Save** your finished setup.

**Right click** on the Layout Tab to **Preview** and **Plot.**

You could also try making multiple Viewports—individually manipulating their size, individually activating the Viewports to position and scale the object(s) displayed, and moving the Deactivated Viewports around your page for the final setup. Make sure that all Viewports are on the Viewport layer so that they won't Plot.

**Save** to the drive location designated by your instructor and create a Backup file

on the Floppy (A:) Drive…per your instructor's direction.

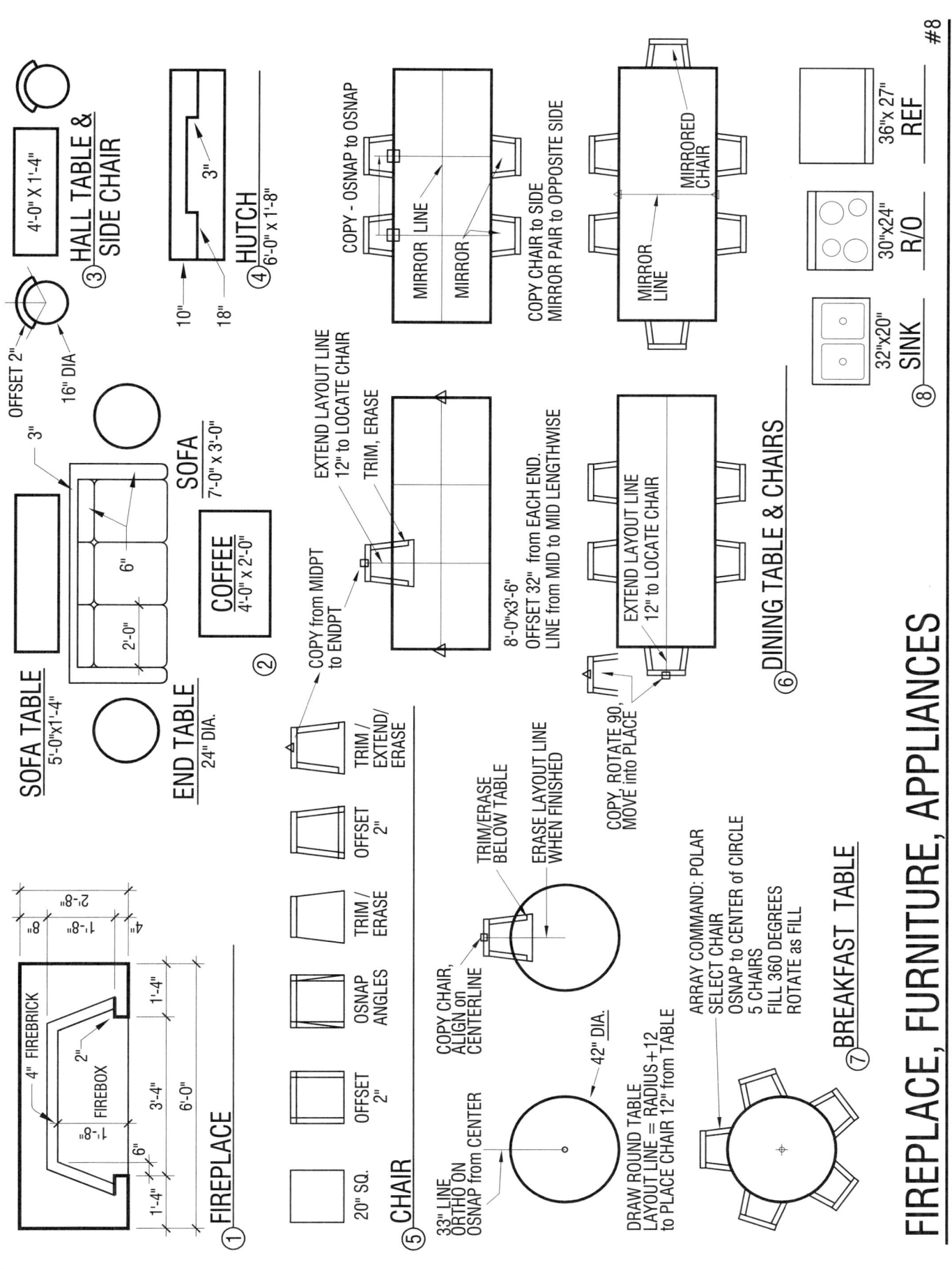

FIREPLACE, FURNITURE, APPLIANCES

#8

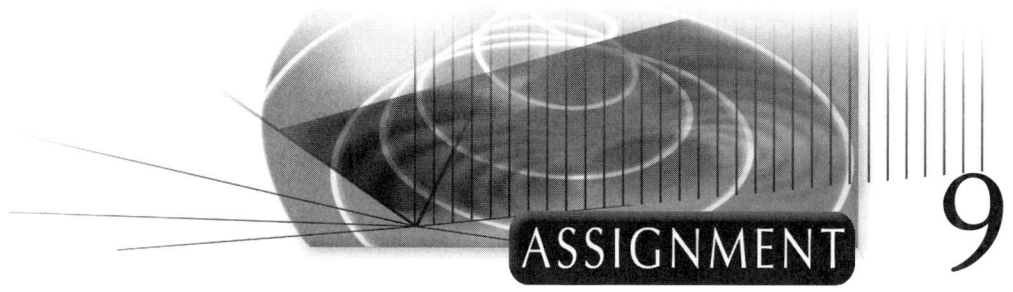

# Floor Plan: Furnishings

## COMPETENCIES/LEARNING OUTCOMES

Upon successful completion of the **Floor Plan: Furnishings,** the student will have used and begun to master the following settings, operations, and drawing commands.

**Settings:**    Units, Limits, Snap, Grid, Polar Tracking, Ortho, OSnap, Layers, Linetypes

**Operations:**  Open/Exit AutoCAD; Access Menus; Enter Commands; Startup Wizard; Open; Save As and Save drawing files; Creating/Saving backup files.

**Commands:**

Line

Offset

Fillet (0)

Erase: Select, Window, Crossing Window

Undo

Zoom Window, Previous, All, Realtime

Pan

Save

Save As

Plot in Paper Space

Trim

OSnap

Fillet with Radius (R) setting

Grips

Circle

Copy and Multiple Copy

Mirror

Layer Properties

Linetype Load/Selection

Linetype Scale

Extend

Properties (Modify and Match)

Arc: SCE (Start, Center, End)

> Array (Rectangular)
>
> Break at Point
>
> Rotate
>
> Array (Polar)

**(new)**  **Move**

**Insert**

**Explode (and Xplode Inherit)**

---

## PROCEDURE

Working alone and with classmates, incorporate information from lectures, demonstrations, and corresponding information in your text to complete this drawing.

This one will be quick and fun—guaranteed to amaze your friends and foes alike. As always…

### RELAX, EXPLORE, ASK QUESTIONS,
### WORK WITH YOUR FELLOW CLASSMATES.

The **Insert** Command: You will be combining some files during this exercise—"Inserting" one drawing into another. The **Insert** command is very valuable—it enables us to store drawings of individual entities (sometimes called Blocks) in a "symbols library," and then **Insert** those details into current projects. For example, if the Kitchen Sink you have drawn was stored in its "own" drawing file, *anytime* in the future you needed a two-compartment Kitchen Sink for a Floor Plan, you could **Insert** this drawing, rather than drawing it from scratch every time you needed it. Now that's what we're talkin' about!

In many offices, each and every "repeatable detail" that is created for a drawing is also stored as an entity in the symbols library for future use. Imagine how the scope of such a library would grow over a period of time, and the time it would save. Offices often purchase software that has many of our standard details drawn and ready to **Insert.** In later exercises you will access such a library and **Insert** Blocks for Water Closets, Lavatories, Sinks, Appliances, etc.

One "glitch" about Inserts…when a drawing file "comes into" your current drawing, it enters that drawing as a BLOCK—that is a "single entity"—like it was "one big continuous line." You cannot Erase single lines, change Colors or Linetypes of individual parts of a BLOCK. If you click one line of the BLOCK in the Erase command, the whole thing will turn to dots and Erase. If you want to perform modifications to a BLOCK, then you must **Explode (or Xplode)** the BLOCK. Once Exploded, the Block's components become individual entities again, and you are off to the races.

You *can,* however, **Move, Copy, Mirror, Rotate,** etc. **BLOCKS** without **Exploding** them.

---

### EXPLODE AND XPLODE

If you want to change some element or property of an inserted drawing or Block—you must Explode the Block (Modify Pulldown or from the Modify Toolbar). In AutoCAD, **when you explode a BLOCK created on Layer 0, it "goes" to Layer 0.** That might mean, depending on how the original drawing was created, that not only has its Layer changed, but its Color/Linetypes as well. If you use Explode, then after doing so, you must Modify Properties of the exploded Block, reassigning the entity to its proper Layer and maybe Colors and/or Linetypes as well. An easy way around this is to **Xplode** rather than **Explode.**

The **Xplode** command will explode the Block into separate entities but allow it to retain its "current" Properties—Layer, Color, Linetype, etc. Type **XPLODE,** Enter, Select Objects, Enter, and then type "I" (for Inherit), Enter. The object returns to separate entities for editing, but will remain on its "current" Layer, with its current Colors, Linetypes, etc.

## BLOCKS, LAYERS, AND COLORS

Watch your Layering when you start **Inserting** objects. Bring objects in on the appropriate Layer, or bring them in and then reassign them to the correct Layer. In some cases you may desire new Layers for the inserted objects. If the objects you're Inserting were originally drawn on Layer 0, then they will automatically land on the Current Layer. If the original objects were drawn to the default Color of Layer 0, then when Inserted, they will default to the Color of the Current Layer as well. This will be the case when you **Insert** your DOOR and WINDOW drawings. If specific Colors were assigned to the original drawing, like you did with your SHUTTER, then those Colors will hold true when you **Insert**. AutoCAD remembers that you wanted those items a "special" Color.

## FLOOR PLAN: FURNISHINGS

**Open** and/or continue working on your \*\*\*GRG drawing (**FLOOR PLAN**)—your settings should already be locked in … **Snap** OFF; **Ortho** OFF; **OSnap** ON; **Fillet** Radius = 0"

You're going to **Insert** your previous work into this drawing, but first, do some rearranging of your screen using the **Move** command.

## MOVE COMMAND

The **Move** Command—new but not new—works just like the Copy command, only the entity "relocates" leaving empty space behind. For accuracy, use OSnaps for Base and Displacement Points, or drag and type to set new locations for the Move command.

**IMORTANT!** If you **Move your Floor plan—or any drawing with multiple Layers—make sure that ALL LAYERS are turned ON**, or you will leave Layers "behind" when you Move. There are ways to fix this, like using OSnaps to Move/Reposition the poor creatures left behind, but it's often a real mess.

**Zoom ALL**

**ALL Layers turned ON**—check the light bulbs in the Layer flyout—ask your instructor if you've doubts.

**Whenever you MOVE an item, especially a drawing like a Floor Plan, which typically has multiple Layers, make sure that ALL your layers are turned ON, so you don't unintentionally leave anything behind in the MOVE.**

**Grid** ON—this will show you your "paper."

**Move** the FLOOR PLAN toward the lower right portion of your screen.

**Move** FIREPLACE, FURNITURE, APPLIANCES to the upper left of your screen.

We will bring our **Inserts** in at the lower left—see your handout for an approximation of this arrangement. It is not necessary that you have things positioned *exactly* like the illustration—just have things in the general quadrant.

**Insert,** each in turn, your drawings of the **DOOR, WINDOW, CHAIRS, SHUTTER,** and **STAIRS**.

Select **Insert/Block** from the **Insert** Pulldown menu at the top of your screen.

Pick **Browse** … the **Select Drawing File** dialog box will appear.

**Look In** … locate the drive on which you've been saving your drawings.

**Select** the drawing you're after … your DOOR first (\*\*\*DR)

Pick **Open** to return to the **Insert** dialog box.

Back at the **Insert** dialog box …

We will want to **Insert** this group of items without changing their size or rotation so …

At **Insertion Point**—**Specify On Screen** box—**select** (check the box).

At **Scale** set *X* and *Y* to a value of **1** (probably is already set at that default).

At **Rotation** set **Angle** to **0** (probably is already set at that default).

Notice that you can **Explode** the drawing automatically by clicking that option in this dialog box. But don't do that here. If you need to **Explode** a drawing, you can do it later with the **Explode** or **Xplode** commands.

Sometimes we want to change the size and/or rotation of a drawing we **Insert.** We can do that directly in the dialog box settings or pick **Specify on Screen** and do it there, as part of the final Insert. This alternate allows us "see" what's happening.

**Pick OK** to leave the dialog box and return to your screen. You will see the inserted drawing "ghosting" around on your screen. Find an empty spot for it to land, and then "pick"—left click—that point. **Make sure that you don't stack one drawing on top of another.**

### PRESTO!

Repeat the process, **Inserting** your DOOR, WINDOW, CHAIRS, SHUTTER, and STAIRS. Your drawing of the FLOOR PLAN should look similar to the boxed example on your handout, but your drawing doesn't have to look exactly like the example. Just make sure that all the pieces are there, somewhere.

## PLACE THE FURNITURE IN THE PLAN

Modify **Properties** command—Before you begin multiplying and spreading your furniture, do a Layer check. Make sure that all furniture—**including the CHAIRS you inserted**—are on the FURNITURE Layer, MAGENTA. Use modify **Properties** to check and/or accomplish this task. Remember that you can grab whole bunches of things to change with a Window or Crossing Window and you can easily change Layers by using the flyout at the Layers Toolbar.

**Assign the DOOR, WINDOW, and SHUTTER** to the ELEVATIONS Layer.

**Assign the STAIRS** to the WALLS_INTERIOR layer.

**Assign CHAIRS** to the FURNITURE Layer.

Notice that Colors you have assigned to these drawings when you created them remain in place as you move them form Layer to Layer (SHUTTER and STAIRS). But Colors for drawings you created on Layer 0 to the default Layer Color (DOOR, WINDOW, and CHAIRS) change "ByLayer" as you move them to different Layers, assuming the default Color for the Layer to which they move.

## PLACE FURNITURE/STAIRS/APPLIANCES/FIREPLACE(S) IN THE PLAN

See example—your Plan may vary, but make sure that you "**Rotate**" a piece or two.

### FURNITURE and STAIRS
Use **Move, Copy, Multiple Copy, Mirror, Rotate, Trim; Zoom** and **Pan** during process.

You will need to **Xplode** the CHAIRS Block to manipulate the Chairs individually. Remember that **Explode** will place entities on Layer 0 and change their Color, but **Xplode**—with the **I** (for Inherit) option—will allow for individual manipulation and also allow the entities to retain their current Properties.

### APPLIANCES
Construct "spaces" in the kitchen countertops to receive the REFRIGERATOR and RANGE/OVEN, a bit larger than reality for the sake of reading your Plot ... try allowing for a one-inch space at each side of the appliances, i.e., "return" the countertop to the wall at each side of these appliances.

### FIREPLACE
You will have to **Rotate** the Fireplace before positioning it. Use OSnaps when you **Move** or **Copy** for Base and Displacement points.

**Copy** the Fireplace to a point next to the plan that will facilitate its Rotation.

**Rotate** the Fireplace 90°.

**Move** the Fireplace into place.

> Select (OSnap) the MIDPOINT of the firebrick lining's face as your Basepoint.
>
> Select (OSnap) the MIDPOINT of the Living Room Wall (outer line) as your Displacement Point. **Trim** out the Fireplace as required.

**Copy** the Fireplace to the Parlor wall using the same **OSnap** procedure to accurately position the unit on the room's axis. If you want, you could **Mirror** another unit into place at the Dining Room, by selecting the Hall Axis as the Mirror Line, OR by **Copying, Rotating,** and **Moving** into "final" position.

**DO NOT ERASE** Door, Window, and Shutter—we will use them later.

You may **Erase** "extra" Furniture and Appliances after the Floor Plan is "furnished."

> **Save** this drawing to your designated location.

## PLOT THE PLAN AND FURNITURE IN PAPER SPACE

Refer to **Quick Guide** as needed.

**Select a Layout Tab** to enter Paper Space:

| | |
|---|---|
| **Page Setup:** | Right click on the Layout Tab for Page Setup. |
| **Plot Device Tab:** | (at top of dialog box) |
| **Plotter Configuration:** | per your instructor |
| **Plot Style Table:** | **monochrome**—select **Edit**—set Lineweights for colors 1–7 per **Quick Guide** for SMALL PLOT. Watch decimal points! Color 7, Black, is the same as White. Don't worry about grays at this time. Select **Save and Close** at the bottom of the dialog box. |
| **Layout Settings Tab:** | (at top of dialog box) |
| **Paper Size:** | letter (8.5 × 11) |
| **Orientation:** | **landscape** |

Other settings are probably OK … your instructor will verify.

**OK** Page Setup and you'll see your designated piece of paper.

Create a **single Viewport** nearly the size of your paper boundaries.

> **Scale** the Viewport to $^1/_8$" = 1'-0"; center the Floor Plan in the Viewport (Pan).

Toggle back to Paper Space and change the Viewport size to include Floor Plan but exclude any portions of the Door, Window, and/or Shutter that may now be appearing.

Make sure that your Viewport is on the VIEWPORT Layer (no Plot).

**Save** your finished setup.

**Right click** on the Layout Tab and select **Plot.**

Double check to make sure that your HIDDEN and CENTERLINES are plotting as such. If not, then **de**select "**use paper space units for scaling**" box in the **Format pulldown/Linetype**

> **Save** to the drive location designated by your instructor and create a Backup file
>
> on the Floppy (A:) Drive per your instructor's direction.

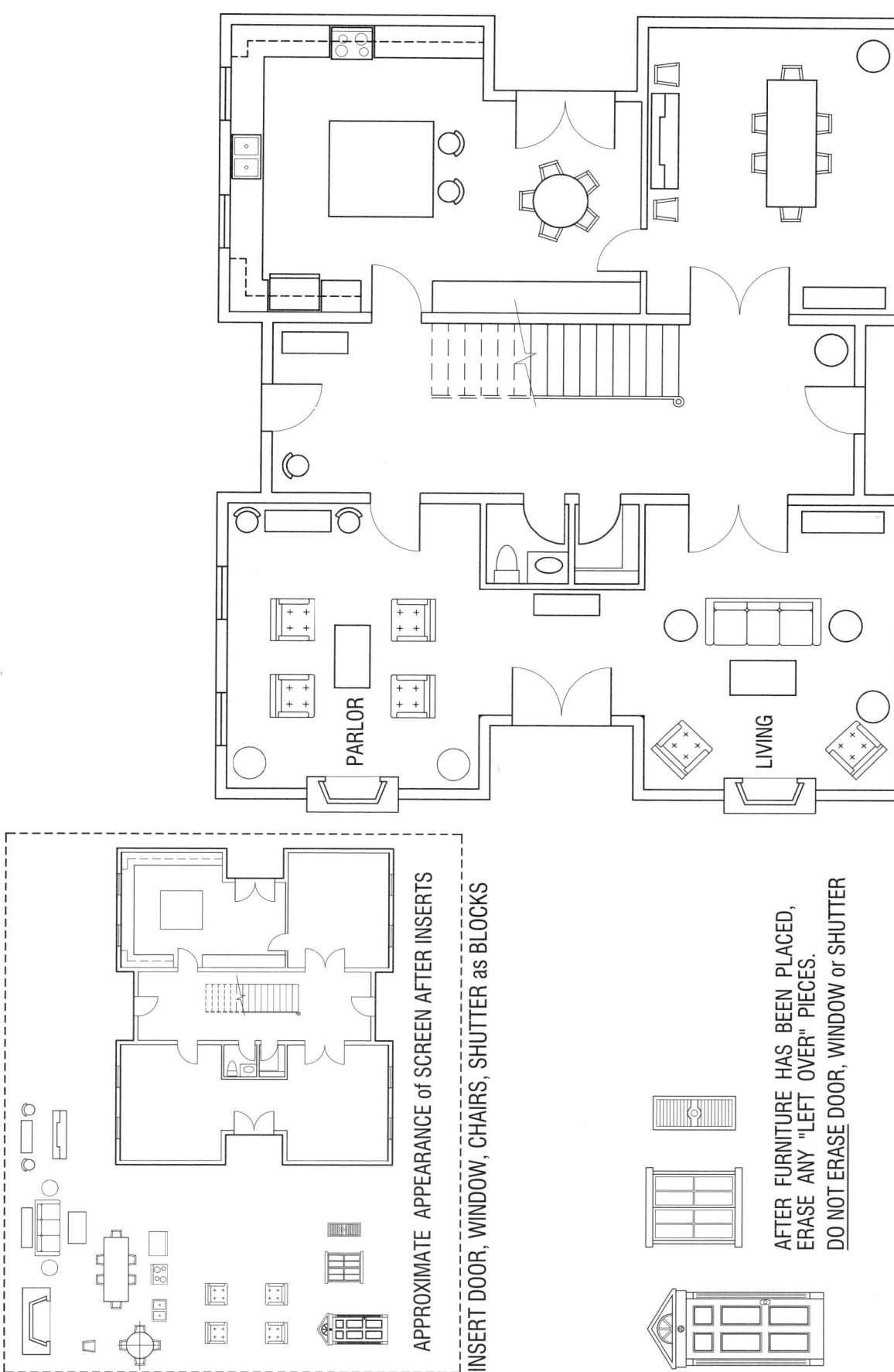

PARLOR

LIVING

APPROXIMATE APPEARANCE of SCREEN AFTER INSERTS

INSERT DOOR, WINDOW, CHAIRS, SHUTTER as BLOCKS

AFTER FURNITURE HAS BEEN PLACED, ERASE ANY "LEFT OVER" PIECES. DO NOT ERASE DOOR, WINDOW or SHUTTER

## GEORGIAN HOUSE: FURNISHINGS

# Floor Plan: Porches and Hearth(s)

## COMPETENCIES/LEARNING OUTCOMES

Upon successful completion of the **Floor Plan: Porches and Hearth(s)**, the student will have used and begun to master the following settings, operations, and drawing commands.

**Settings:** Units, Limits, Snap, Grid, Polar Tracking, Ortho, OSnap, Layers, Linetypes

**Operations:** Open/Exit AutoCAD; Access Menus; Enter Commands; Startup Wizard; Open; Save As and Save drawing files; Creating/Saving backup files.

**Commands:**

Line

Offset

Fillet (0)

Erase: Select, Window, Crossing Window

Undo

Zoom Window, Previous, All, Realtime

Pan

Save

Save As

Plot in Paper Space

Trim

OSnap

Fillet with Radius (R) setting

Grips

Circle

Copy and Multiple Copy

Mirror

Layer Properties

Linetype Load/Selection

Linetype Scale

Extend

Properties (Modify and Match)

Arc: SCE (Start, Center, End)

Array (Rectangular)

Break at Point

Rotate

Array (Polar)

Move

Insert

Explode (and XPLODE Inherit)

## PROCEDURE

Working alone and with classmates, incorporate information from lectures, demonstrations, and corresponding information in your text to complete this drawing.

**RELAX, EXPLORE, ASK QUESTIONS,**

**WORK WITH YOUR FELLOW CLASSMATES.**

## PORCHES

**Open** and/or continue working on your \*\*\*GRG drawing (**FLOOR PLAN**)—your settings should already be locked in … **Snap** OFF; **Ortho** ON; **OSnap** ON; **Fillet** Radius = 0"

1. Make WALLS_EXTERIOR Layer Current.

   Using the **Offset, Fillet, Trim, Mirror,** etc. commands, draw the four PORCHES per handout.

2. **Modify Properties:** Change the Color of the *Porch steps only* to BLUE.

## HEARTH

On the WALLS_EXTERIOR Layer

1. Draw a HEARTH extending 1'-8" (minimum) × 6' in front of each Fireplace.

2. **Modify Properties:** Change the Color of the Hearth(s) to BLUE.

**SAVE** this drawing to your designated drive.

## PLOT THE FLOOR PLAN WITH PORCHES

Refer to **Quick Guide** as needed.

**Select a Layout Tab** to enter Paper Space:

| | |
|---|---|
| **Page Setup:** | Right click on the Layout Tab for Page Setup. |
| **Plot Device Tab:** | (at top of dialog box) |
| **Plotter Configuration:** | per your instructor |
| **Plot Style Table:** | **monochrome**—select **Edit**—set Lineweights for colors 1–7 per **Quick Guide** for SMALL PLOT. Watch decimal points! Color 7, Black, is the same as White. Don't worry about grays at this time. Select **Save and Close** at the bottom of the dialog box. |
| **Layout Settings Tab:** | (at top of dialog box) |
| **Paper Size:** | letter (8.5 × 11) |
| **Orientation:** | **landscape** |

Other settings are probably OK…your instructor will verify.

**OK** Page Setup and you'll see your designated piece of paper.

Create a **single Viewport** nearly the size of your paper boundaries.

**Scale** the Viewport to $^1/_8$" = 1'-0"; center the FLOOR PLAN (only) in the Viewport (**Pan**). (Some DOOR, WINDOW, SHUTTER may be peeking in)

Double Click or Toggle back to **Paper Space** and change the Viewport size (Grips) to include FLOOR PLAN but exclude any portions of the DOOR, WINDOW, and/or SHUTTER that may now be appearing.

Make sure that your Viewport is on the VIEWPORT Layer (no Plot).

**Save** your finished setup.

**Right click** on the Layout Tab to **Preview** and **Plot**.

Double check to make sure that your HIDDEN and CENTERLINES are plotting as such. If not, then **de**select "**Use paper space units for scaling**" box in the **Format Pulldown/Linetype.**

**Save** to the drive location designated by your instructor and create a Backup file

on the (A:) Drive…per your instructor's direction.

8" REVEAL

12" TREADS

2'-0"

5'-4"

1'-8" HEARTH

10'-0"

2'-8"

2'-8" SQUARE

12" TREADS

4" REVEAL

NOTE: PORCHES WILL MIRROR to the
OPPOSITE SIDES of BUILDING
(USE BUILDING AXES as "MIRROR LINES")

# GEORGIAN HOUSE: PORCHES and HEARTHS

#10

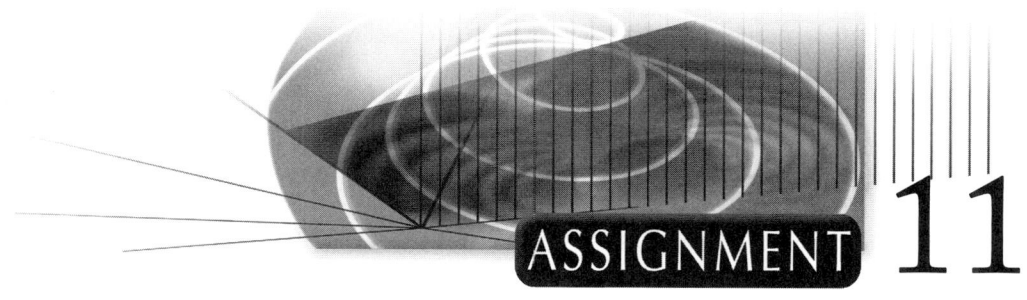

# Floor Plan:
# Text

## COMPETENCIES/LEARNING OUTCOMES

Upon successful completion of the **Floor Plan: Text,** the student will have used and begun to master the following settings, operations, and drawing commands.

**Settings:**    Units, Limits, Snap, Grid, Polar Tracking, Ortho, OSnap, Layers, Linetypes, **Text Style**

**Operations:**    Open/Exit AutoCAD; Access Menus; Enter Commands; Startup Wizard; Open; Save As and Save drawing files; Creating/Saving backup files.

**Commands:**

Line

Offset

Fillet (0)

Erase: Select, Window, Crossing Window

Undo

Zoom Window, Previous, All, Realtime

Pan

Save

Save As

Plot in Paper Space

Trim

OSnap

Fillet with Radius (R) setting

Grips

Circle

Copy and Multiple Copy

Mirror

Layer Properties

Linetype Load/Selection

Linetype Scale

Extend

Properties (Modify and Match)

Arc: SCE (Start, Center, End)

Array (Rectangular)

Break at Point

Rotate

Array (Polar)

Move

Insert

Explode (and Xplode Inherit)

**(new)   Text Style, Text, DT (DText/Dynamic Text), Edit Text (ED)**

## PROCEDURE

Working alone and with classmates, incorporate information from lectures, demonstrations, and corresponding information in your text to complete this assignment.

Before you start placing **Text** (notes, labels, titles, etc.) in a drawing, you must "set" a **Text Style.** If you do not create a **Text Style** with a specific **Font,** AutoCAD will default to its *Standard Text Style* with its default *Font.* The standard font is not typically considered to be representative of Architectural lettering styles. This exercise will direct you to select a Font called Stylus BT—verify the preferred Font(s) with your instructor. Preferences will vary from instructor to instructor, office to office.

As always…

### RELAX, EXPLORE, ASK QUESTIONS,

### WORK WITH YOUR FELLOW CLASSMATES.

**Open** and/or Continue working on your ***GRG drawing (**FLOOR PLAN**)

TEXT: Setting a Style/Font:

1. From **Format** Pulldown, **select Text Style.** The **Text Style** dialog box will appear.

2. Pick **New,** then type over *Style 1* with the name **ARCHITECTURAL.** Pick **OK.**

3. At **Font Name** box, **scroll to** and **select Stylus BT**—verify with your instructor.

   A sample of the Font will appear in the **Preview** window at the lower right.

4. At **Height:** *leave the Height set at 0"*—verify with your instructor.

5. In the **Effects** box, set **Width Factor** at .8—verify with your instructor.

6. **Pick Apply** to "set" the new style/font.

7. **Pick Close** to exit the dialog box.

You have now selected a Font (**Stylus BT**) and named it as your "ARCHITECTURAL" Style. This setting will remain part of your drawing file—but for a New drawing, you must again set a **Text Style** to avoid AutoCAD's default Font.

---

### NAMING TEXT STYLE

Text Style and Font are connected, but are different. What you "do" with a Font —set a height, for instance—produces a specific Style. That is to say, you could have the *same* Font for a Text Style that's 10" high and one that's 5" high. It's tempting to alter the Font without giving a new name to the Style, but doing so changes AutoCAD's default Font for its Standard Text Style. Such Text, when tied to Inserted Blocks, will revert to the Font of the current Standard Style and sometimes that's problematic. For this and other reasons, it's probably best to give each "new" Font selection a name, even if everything else about them is the same.

## TEXT HEIGHTS

AutoCAD will allow you to set individual Text heights for separate Text Styles or to set heights as you apply Text to your drawing. If you needed two different heights for a drawing, you could name two separate Text Styles, each with the same Font but with different heights set in the Text Style dialog box. Then before applying the Text to a drawing, you either select the Style you need from the Styles Toolbar or you go to the Text Style dialog box, select the Style you need, pick Apply, and you're good to go.

You were asked, however, to leave the height in the dialog box at 0". This will allow you to vary the Text height as you apply it to the drawing. In one way, an extra step, in another way a more direct approach. Setting height(s) in the dialog box will override other Text settings, including Dimension Text, and typically prove to be troublesome…unless you decide to set a verity of Styles with different heights and then "switch" back and forth among the styles.

### TEXT: Placing Text on the Floor Plan

You've just set a **Text Style**—so you're ready to go. Remember that with each New drawing (unless you are using a template), to avoid the default settings, you must set a Text Style and Font.

You have several methods for placing text on your drawing including **Single** and **Multiline Text** options in the **Draw** Pulldown menu, selecting the **Multiline Text** icon (Draw Toolbar), or typing **DT** (Dynamic Text).

Try placing some text using all four of these options. This exercise will suggest the **DT** option because with that option you can "see" the text appearing on the screen as you type.

## TEXT HEIGHTS IN MODEL SPACE—SEE SETTING TEXT HEIGHTS SUPPLEMENT

AutoCAD Plots Text placed in Model Space to scale, just like any other entity placed in Model Space. This typically means that you must consider both the scale of the final Plot and the desired Plotted size of the Text before deciding Text height. For example, at $1/4" = 1'-0"$ scale, 12" high letters will Plot $1/4"$ high, 6" high letters will Plot $1/8"$ high. Though we will complete some check Plots at smaller scales, Text heights will be gauged for a $1/4" = 1'-0"$ Plot. Refer to **Supplement 3: Setting Text Heights in Model Space** for more information.

### Make TEXT Layer Current
**"CAPS LOCK" ON** … Architectural lettering is typically upper case.

### PLACING TEXT: TWO WAYS

Place room names and notes on the Floor Plan as shown. Though we are Plotting $1/8" = 1'-0"$ "check Plots," select Text heights for a $1/4" = 1'-0"$ Plot.

### ONE WAY: SINGLE LINE TEXT

This method displays the text at the command prompt line AND directly on your drawing as you type. Follow the command line prompts:

> Type **DT** (for "single line" Dynamic Text) then **Enter** …
>
> > or from **Draw Pulldown/Text/Single Line** (same as typing **DT**)
>
> Pick the **starting point.**
>
> Designate/Enter the desired **Height**—refer to **Supplement 3: … Text Heights** …
>
> > For $1/4"$ Plots, try 9" for titles, 6" for room names, 4.5" for typical notes.

Just **Right Click** or **Enter** if you "like" the default setting displayed in brackets.

Once adjusted, the new setting will default until you change it again.

Designate/Enter the desired **Rotation** angle.

0° = horizontal text, 90° will orientate text to the right-hand side of drawing.

**Type the text.** Backspace to correct errors.

**Enter once** for a second line of text.

**Enter twice** to end the command.

## ANOTHER WAY: MULTILINE TEXT (SAME AS PICKING THE MULTILINE TEXT ICON)

In this mode you must select a window for the text. A "screen" will then appear, which allows you to set text height (as well as other options). Locate the *upper left* corner of the window where you want your text to "begin"—you can "twink" this starting point by sliding the margin adjustment before typing.

From the **Draw** Pulldown

Pick **Text/Multiline** (or Pick the **Multiline Text Icon**).

Pick **Text Window** location.

Enter/Verify Height for your text—see **Supplement 3: … Text Heights …**

For ¼" Plots try 9" for titles, 6" for room names, 4.5" for typical notes.

Type the Text—backspace to correct errors.

**OK** to finish.

---

### SINGLE LINE OR MULTILINE?

First of all Single Line Text commands allow you to type more than one line—so don't be deceived by the name. Many think that the Single Line option (DT) offers more immediate control over Text location while the Multiline option readily presents more editing controls and is probably favorable for "long" notes. In Single Line application, each line is an entity; in Multiline Text, the entire paragraph(s) is a single entity.

---

## EDITING TEXT: THREE WAYS
### ONE WAY TO EDIT TEXT: TYPE ED (FOR EDIT) AND ENTER

**Select Text** and make corrections. This works for Dimensions too!

If you created your text as Multiline, then the Multiline Text Formatting window will appear, giving you content and formatting options. If you created your text as Single Line, then this route will only allow you to edit content. To change heights and other formatting of Single Line Text, you must either use the **Modify Properties** command, or **convert the text to Multiline** using **Express** (pulldown) tools.

### ANOTHER WAY TO EDIT TEXT: MODIFY PROPERTIES

From **Modify/Properties** pulldown or **Icon**.

**Select Text** to edit and make any of a number of changes, including changes to height, rotation, and/or Font.

### ANOTHER WAY TO EDIT TEXT: EXPRESS PULLDOWN—FOLLOW PROMPTS

**Try all three for the experience.** Text Editing is a valuable command to have in your armory. Remember that Text can **Move, Copy, Multiple Copy, Rotate, Erase,** etc., like any other entity. Also you can **Move Text** using **Grips** and **Match Properties** (paint brush) between Text Styles.

## PLOT THE FLOOR PLAN WITH TEXT IN PAPER SPACE

Refer to **Quick Guide** as needed.

**Select a Layout Tab** to enter Paper Space:

| | |
|---|---|
| **Page Setup:** | Right click on the Layout Tab for Page Setup. |
| **Plot Device Tab:** | (at top of dialog box) |
| **Plotter Configuration:** | per your instructor |
| **Plot Style table:** | **monochrome**—select **Edit**—set Lineweights for colors 1–7 per **Quick Guide** for SMALL PLOT. Watch decimal points! Color 7, Black, is the same as White. Don't worry about grays at this time. Select **Save and Close** at the bottom of the dialog box. |
| **Layout Settings Tab:** | (at top of dialog box) |
| **Paper Size:** | letter (8.5 × 11) |
| **Orientation:** | **landscape** |

Other settings are probably OK… your instructor will verify.

**OK** Page Setup and you'll see your designated piece of paper.

Create a **single Viewport** nearly the size of your paper boundaries.

    **Scale** the Viewport to $^1/_8'' = 1'\text{-}0''$; center the FLOOR PLAN (only) in the Viewport (**Pan**).

        (Some DOOR, WINDOW, SHUTTER may be peeking in)

Double Click or Toggle back to **Paper Space** and change the Viewport size (Grips) to include FLOOR PLAN but exclude any portions of the DOOR, WINDOW, and/or SHUTTER that may now be appearing.

Make sure that your Viewport is on the VIEWPORT Layer (no Plot)

**Save** your finished setup.

**Right click** on the Layout Tab to **Preview** and **Plot.**

Double check to make sure that your HIDDEN LINES, CENTERLINES, etc., are plotting as such. If not, then **de**select "**Use paper space units for scaling**" box in the **Format Pulldown/Linetype.**

    **Save** to the drive location designated by your instructor and create a Backup file

        on the Floppy Drive A:… per your instructor's direction.

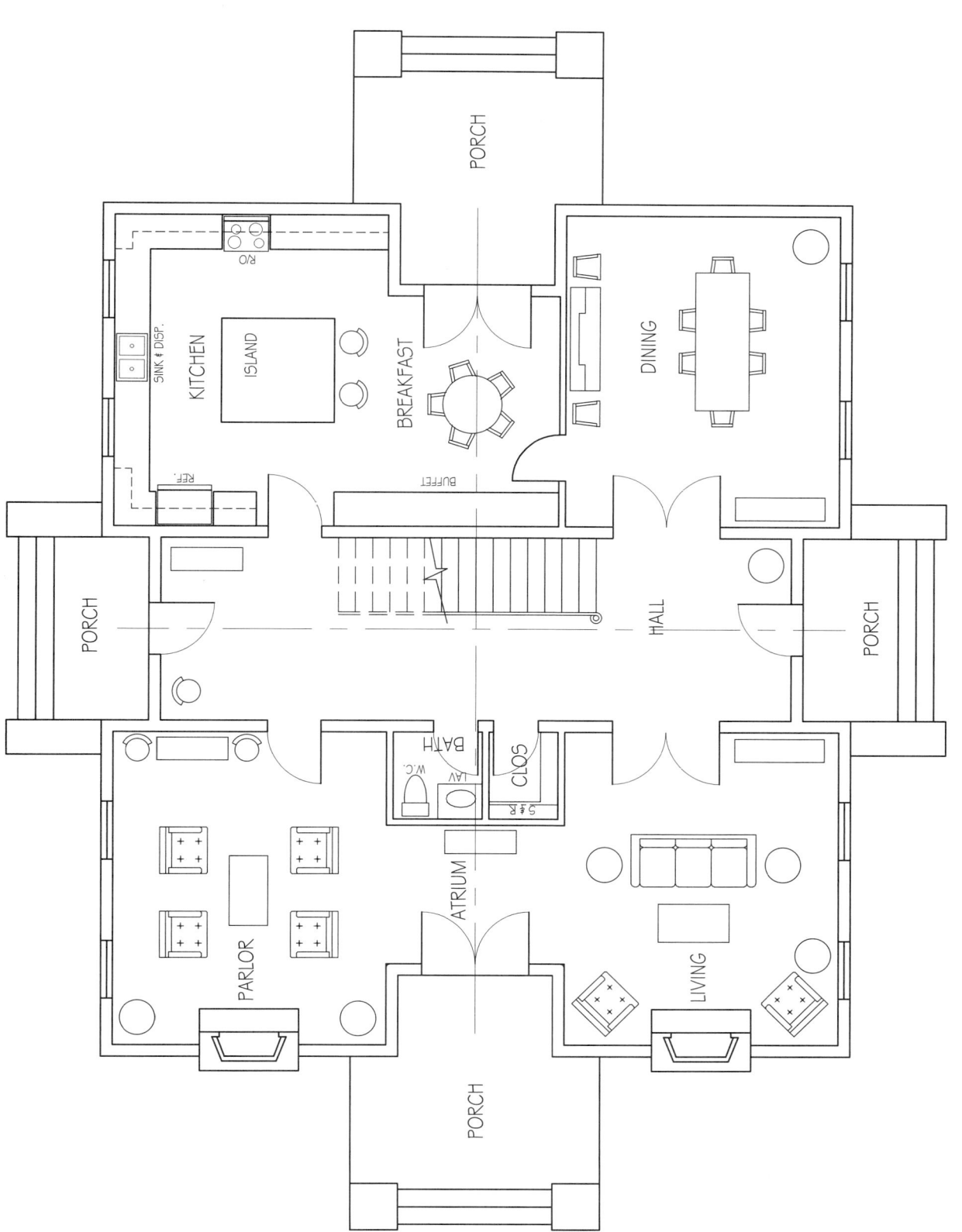

GEORGIAN HOUSE: TEXT

#11

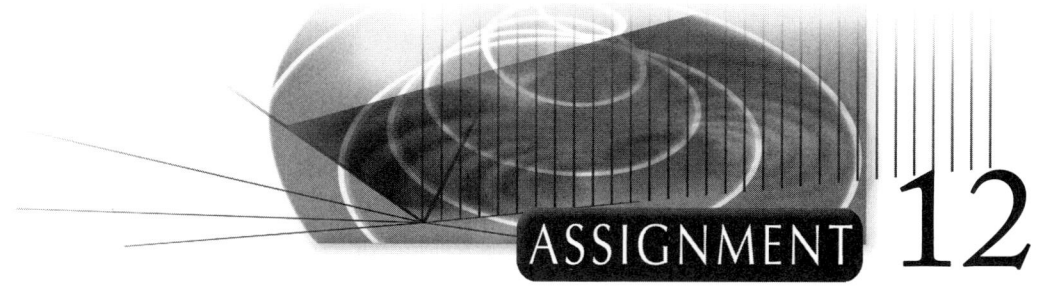

# Dimensions

---

## COMPETENCIES/LEARNING OUTCOMES

Upon successful completion of the **Dimensions,** the student will have used and begun to master the following settings, operations, and drawing commands.

**Settings:** Units, Limits, Snap, Grid, Polar Tracking, Ortho, OSnap, Layers, Linetypes, Text Style, **Dimension Style**

**Operations:** Open/Exit AutoCAD; Access Menus; Enter Commands; Startup Wizard; Open; Save As and Save drawing files; Creating/Saving backup files.

**Commands:**

Line

Offset

Fillet (0)

Erase: Select, Window, Crossing Window

Undo

Zoom Window, Previous, All, Realtime

Pan

Save

Save As

Plot in Paper Space

Trim

OSnap

Fillet with Radius (R) setting

Grips

Circle

Copy and Multiple Copy

Mirror

Layer Properties

Linetype Load/Selection

Linetype Scale

Extend

Properties (Modify and Match)

Arc: SCE (Start, Center, End)

Array (Rectangular)

Break at Point

Rotate

Array (Polar)

Move

Insert

Explode (and Xplode Inherit)

Text Style, Text, DT (DText/Dynamic Text), Edit Text (ED)

**(new)    Dimension**

## PROCEDURE

Working alone and with classmates, incorporate information from lectures, demonstrations, and corresponding information in your text to complete this assignment.

As was the case for **Text,** you will be setting a **Dimension Style** prior to dimensioning the Floor Plan. This process can be a bit daunting, especially to those who are new to architectural drafting and dimensioning formats. Be sure to ask questions if you get stumped and/or have your instructor verify your settings before you proceed.

The good news is that once a **Dimension Style** is set, the actual dimensioning is as close to automatic as it gets—it's a REAL KICK! So hang tight and try not to let too much blood roll out your ears. In advanced classes you will learn ways to copy these settings from one drawing to another using the **Design Center** and/or ways of using Templates with the settings "already there." In the meantime, while you're learning, it's more tough love here to put you through the paces for these types of setups.

You will notice from the example that the goal of this assignment is not to completely dimension the Floor Plan, but to introduce you to the basic concepts of setting a **Dimension Style** and applying what you will learn to be the **Linear** and **Continue Dimension** commands. You may, of course, at your instructor's direction, go beyond the scope shown in the example.

**RELAX, EXPLORE, ASK QUESTIONS,**

**WORK WITH YOUR FELLOW CLASSMATES.**

## ARCHITECTURAL DIMENSIONING FORMATS

Dimensions are used to describe SIZE and LOCATION, two distinct but equally important tasks. We need to tell the builder "how big something is" *and* "where that thing goes."

Some terms to get started—review the drawing for this unit to see the **Dimension Style** we're after. In typical architectural formats, the DIMENSION TEXT lists the required measurement in feet and inches and is typically "centered on" and "above" the DIMENSION LINE. The DIMENSION LINE itself shows the extents of the DIMENSION, being terminated by TICK MARKS (not arrows) at each end. EXTENSION LINES "extend" from the drawing to the DIMENSION LINE. The intersection of the DIMENSION LINE and the EXTENSION LINE is marked by the TICK.

Again, review the **Dimension Style** shown on the example drawing. Identify specific features that are common to architectural dimensioning styles, including the use of *ticks* instead of arrows, Extension Lines *offset* from corners of the building, a *crossing* of Dimension and Extension Lines at the tick mark, Dimension Text *aligned* with the Dimension Lines (reading from the bottom or the right-hand side of the drawing), Dimensions *above* the Dimension Lines, the use of architectural units (feet and inches), a $^3/_{32}$ to $^1/_8$" high "architectural" font, and the expression of 0" for "inchless" dimensions.

## SET TEXT STYLE FIRST

Before setting a **Dimension Style(s),** set a **Text Style(s).** You will be asked to designate a Text Style for your Dimension Text, so having a Text Style previously set saves some backtracking. Set the Text Style with a default height of 0" so that setting doesn't override the height setting for the Dimension Text (unless you're planning to use a specific Text Style for Dimensioning only). You have already set your **Text Style** to **Stylus BT** with the 0" default height (per your previous handout) so you're good to go, since you will most probably want to select the same Font for the Dimension Text as you selected for the general drawing Text.

Finally, although there is general agreement on architectural formats, your instructor may very well have some alternate suggestions for some of the settings.

**So … get things ready for setting a Dimension Style and Dimensioning your drawing …**

**Open** and/or continue working on your \*\*\*GRG drawing (FLOOR PLAN).

**Reset** your **Units** from the **Format pulldown/Units.**

> **Select Units**
>
> **Reset** Precision to $^1/_{32}$". Everything else remains the same. **OK.**

**Make DIMENSION Layer "Current"** … and you're off!

## SETTING A DIMENSION STYLE

1. From **Dimension** Pulldown … **select Style … or …**

   from **Format** Pulldown … **select Dimension Style**

   the **Dimension Style Manager** dialog box appears.

2. Pick **New** and the **Create New Dimension Style** box appears.

   Enter **New** name as **ARCHITECTURAL 48.**

   Pick **Continue.**

**All Right!—now you're "stylin' "!**

**From the tabs** at the top of the **Dimension Style Manager** dialog box …

3. **LINES AND ARROWS TAB:**

| | |
|---|---|
| DIMENSION LINES | **COLOR = ByLayer** {Dim Layer is RED (thin)} |
| | **LINEWEIGHT = BY BLOCK** |
| | **EXTEND BEYOND** … (we'll set this later) |
| | **BASELINE =** $^3/_8$**"** |
| EXTENSION LINES | **COLOR = ByLayer** |
| | **LINEWEIGHT = BY BLOCK** |
| | **EXTEND BEYOND =** $^1/_{16}$**"** |
| | **OFFSET ORGIN =** $^1/_{16}$**"** |
| ARROWHEADS | **ARCHITECTURAL TIC (first and second)** |
| | **ARROW SIZE =** $^1/_{16}$**"** |
| DIMENSION LINES | **EXTEND BEYOND =** $^1/_{16}$**"** |

   … had to "come back" to this one, couldn't set it until you set

   Architectural Tics as the arrowhead style

4. **TEXT TAB:**

| | |
|---|---|
| APPEARANCE | **STYLE = Architectural** |

You can do this ONLY because you have already set the "Architectural" Text Style—last unit . . . right?

|  | **COLOR = GREEN** |
|---|---|
|  | **HEIGHT =** $3/32$**"** (might have to "type" this guy) |
| PLACEMENT | **VERTICAL = ABOVE** |
|  | **HORIZONTAL = CENTERED** |
|  | **OFFSET =** $1/32$**"** |
| ALIGNMENT | **WITH DIMENSION LINE** |

## 5. FIT TAB:

OPTIONS    **ALWAYS KEEP BETWEEN LINES**

PLACEMENT  **OVER LINE WITHOUT LEADER**

SCALE for    **USE OVERALL SCALE OF**—enter **48** (Scale Factor for $1/4$" Plot)

    DIM. FEATURES

This value (48) is the so-called Scale Factor for the scale of the final Plot. See *Supplement 2: Drawing Area/Limits* . . . for scale factors for a variety of architectural scales. If you intended to plot this drawing at $3/8$" = 1'-0", you would use a value of 32 here. For a $1/2$" Plot, a value of 12. It's this scale factor that we tagged on to our Dimension Style Name—Architectural 48 . . . remember?

FINE TUNE    **ALWAYS DRAW DIMENSION LINE**

## 6. PRIMARY UNITS TAB:

| LINEAR DIMENSIONS | **UNIT FORMAT = ARCHITECTURAL** |
|---|---|
|  | **PRECISION =** $1/16$**"** |
|  | **FRACTIONS = Horizontal** |
| MEASUREMENT SCALE | **SHOULD BE SET TO 1.000** |
| ZERO SUPPRESION = | **0 Feet—ON (checked)** |
| = | **0 Inches—OFF (not checked)** |

## 7. ALTERNATE UNITS and TOLERENCES TABS?

GREAT NEWS!—FORGET'EM FOR NOW.

## 8. **OK** at the bottom of the dialog box—takes you back to the **Style Manager** dialog box.

Pick **Set Current**—this makes the style you just created the "current style" in use. You can and will create other dimension styles in the future, and you will then "pick" the appropriate one as the "current" style.

Notice the **Modify** option at the right—by selecting this option, you can adjust these settings if you need to make a change in the future. The adjustment, if completed to the "current" style, will automatically update the existing dimensions on your drawing.

## 9. **Close** and you'rrrrrrre DONE! WHEW! Are we talking *settings* or what?!

**The Modify option in the Dimension Style Manager will let you adjust these settings if needed.**

While a variety of ways exist that will allow you to avoid having to set **Text** and **Dimension Styles** with each new drawing, you will be asked to repeat both operations one more time for the next group of drawings. Practice like this begins to give you deeper understanding of AutoCAD's workings as well as an appreciation for the shortcuts to come.

## DIMENSIONING A DRAWING

The purpose of this dimensioning exercise is to explore methods of simple **Linear** dimensioning systems on AutoCAD. You are not required to show more dimensions than indicated in the attached handout, but you most certainly may go beyond the handout if you desire. Although there are methods for automatically spacing Dimension Lines, in these introductory assignments you will "eyeball" their locations. Try to hold dimension locations "close" to the positions shown on your handout—that is, don't let the dimension lines stray too far away from the drawing. Verify the Dimension Line location with your instructor.

---

### MANIPULATING DIMENSION TEXT, DIMENSION, AND EXTENSION LINES

Once in place, Dimension Line, Extension Line, and Dimension Text can be adjusted using Grips. Just Pick them up and Set them down. Dimension Text can be edited through the **ED** command—be careful about this; it can lead to big trouble.

Entire strings of Dimensions may be moved closer or further from their target using the **Stretch** Command. You MUST pick the Dimensions with a Crossing Window for this to work.

If a Dimension shows as incorrect, first check the accuracy of your OSnap at the Extension Line origin before you flip your lid. Zoom IN tightly—you may have likely missed your target.

---

**Using your OSnaps is essential for accuracy! Using your OSnaps is essential for accuracy!**

**Using your OSnaps is essential for accuracy!** ......................................................Got that?

You may also wish to dock the **Dimension Toolbar** (if it is not already displayed on your screen) by Right Clicking on any Toolbar and picking the Dimension Toolbar—or you can work from the **Dimension** Pulldown menu.

**For each string,** the first Dimension command will be **Linear,** subsequent dimensions **on the same string** will fall to the **Continue** command. You can **Zoom** IN and OUT during this process for accuracy.

Ok, here ya' go! **Make sure that your Dimension Layer is Current.**

1. From the **Dimension** Pulldown—or **Toolbar**

   Pick **Linear**

2. You are prompted for the "**First Origin**"

   —OSnap to the target corner of one Extension line.

   You are prompted for the "**Second Origin**"

   —OSnap to the opposite Extension line.

   You are prompted for the "**Dimension Location**"

   —"eyeball" a distance below the house similar to the example and "Pick"

### PRESTO! INSTANT DIMENSION!

3. To "continue" a string, select **Continue** from the Dimension Toolbar or Pulldown

   You are prompted only for the "**Second**" origin.

   —Pick the "next target" on your dimension string,

   —And just keep picking to finish the string. Enter to stop.

4. Repeat the entire process beginning with **Linear** and following with **Continue** for the remaining strings as shown.

If the dimension shown isn't what it is supposed to be—don't blame the computer, blame the person who drew the building (we won't mention any names if you won't).

One reason the dimension may not be showing correctly is that you've snapped to the wrong point. As stated earlier, you can manipulate Extension line, Dimension line, and Text locations using Grips. If you relocate the Extension line, the Dimension text will automatically adjust—now that's gotta' put a smile on your face!

You can **Edit** the text without having to "redraw" our building, but that action starts you down a slippery slope. Check with your instructor if you end up with "incorrect" dimensions.

---

### CONTINUE DIMENSION

If you find mysterious 0" Dimensions showing up on your drawing, that means that you're probably giving your Continue Dimensions an "extra" click. Remember that you don't have to select a starting point for a Continued Dimension—they are already tagged into that point.

The Continue command automatically locks onto the last Dimension placed and heads toward the direction in which the original Dimension was placed. You may Select an alternate Dimension source to Continue by typing S (for Select) and Picking the Dimension you wish to Continue.

---

**Save** this drawing to your designated drive.

---

### PLOT THE FLOOR PLAN WITH DIMENSIONS

Refer to **Quick Guide** as needed.

**Select a Layout Tab** to enter Paper Space:

| | |
|---|---|
| **Page Setup:** | Right click on the Layout Tab for Page Setup. |
| **Plot Device Tab:** | (at top of dialog box) |
| **Plotter Configuration:** | per your instructor |
| **Plot Style Table:** | monochrome—select **Edit**—set Lineweights for colors 1–7 per **Quick Guide** for SMALL PLOT. Watch decimal points! Color 7, Black, is the same as White. Don't worry about grays at this time. Select **Save and Close** at the bottom of the dialog box. |
| **Layout Settings Tab:** | (at top of dialog box) |
| **Paper Size:** | letter (8.5 × 11) |
| **Orientation:** | landscape |

Other settings are probably OK . . . your instructor will verify.

**OK** Page Setup and you'll see your designated piece of paper.

Create a **single Viewport** nearly the size of your paper boundaries.

    **Scale** the Viewport to $1/8$" = 1'-0"; center the FLOOR PLAN (only) in the Viewport (**Pan**).

    (Some DOOR, WINDOW, SHUTTER may be peeking in)

Double Click or Toggle back to **Paper Space** and change Viewport size (Grips) to include FLOOR PLAN but exclude any portions of the DOOR, WINDOW, and/or SHUTTER that may now be appearing.

Make sure that your Viewport is on the VIEWPORT Layer (no Plot)

**Save** your finished setup.

**Right click** on the Layout Tab to **Preview** and **Plot.**

Double check to make sure that your HIDDEN LINES, CENTERLINES, etc., are plotting as such. If not, then **de**select "**Use paper space units for scaling**" box in the **Format Pulldown/Linetype.**

    **Save** to the drive location designated by your instructor and create a Backup file

        on the Floppy Drive (A:) . . . per your instructor's direction.

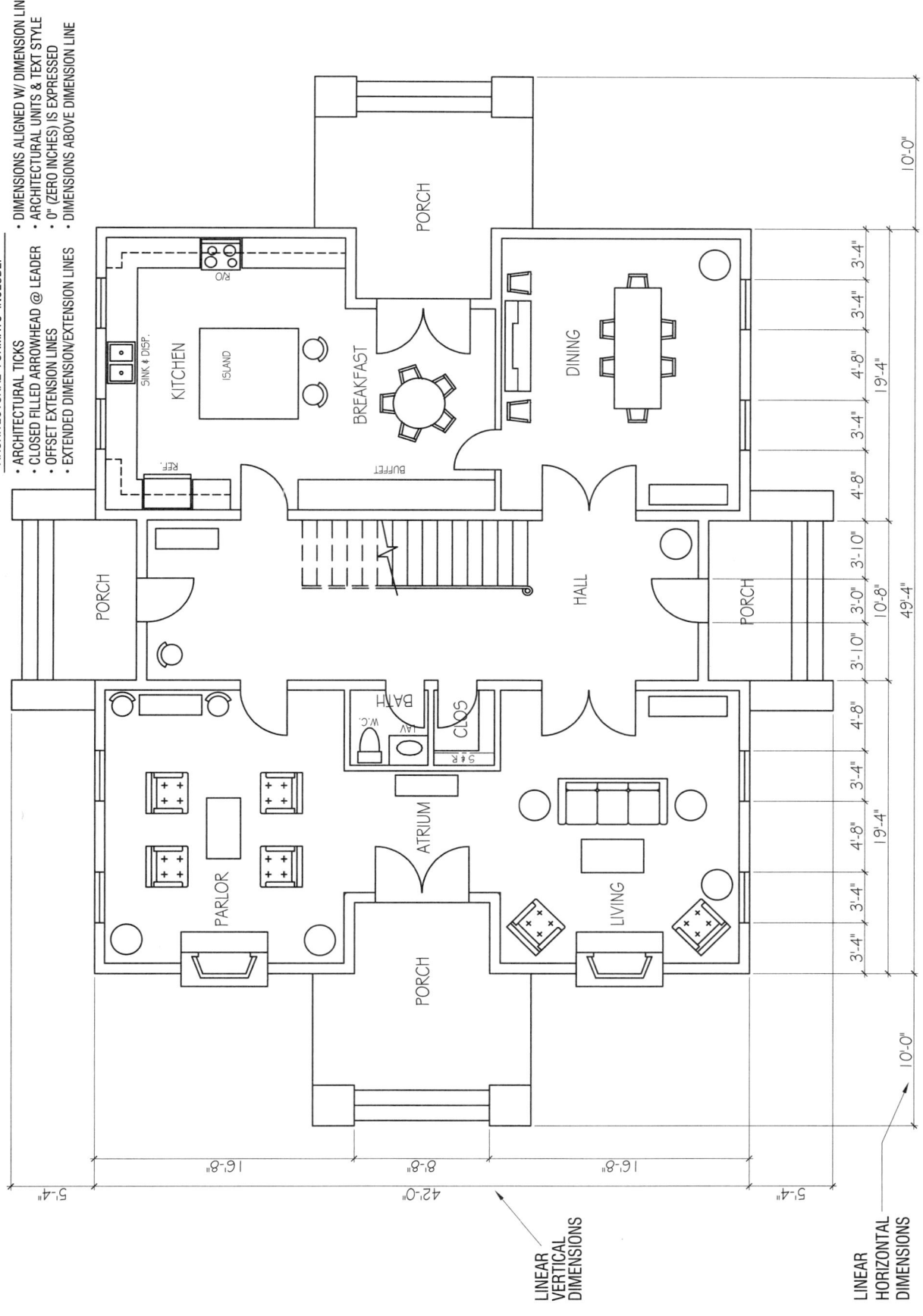

ARCHITECTURAL FORMATS INCLUDE:

- ARCHITECTURAL TICKS
- CLOSED FILLED ARROWHEAD @ LEADER
- OFFSET EXTENSION LINES
- EXTENDED DIMENSION/EXTENSION LINES
- DIMENSIONS ALIGNED W/ DIMENSION LINES
- ARCHITECTURAL UNITS & TEXT STYLE
- 0" (ZERO INCHES) IS EXPRESSED
- DIMENSIONS ABOVE DIMENSION LINE

PORCH

KITCHEN

SINK & DISP.

ISLAND

R/O

BREAKFAST

REF.

BUFFET

DINING

PORCH

HALL

PORCH

BATH
W.C.
LAV
CLOS
5 & R

PARLOR

ATRIUM

LIVING

PORCH

LINEAR
VERTICAL
DIMENSIONS

LINEAR
HORIZONTAL
DIMENSIONS

10'-0"
3'-4"
3'-4"
19'-4"
4'-8"
3'-4"
4'-8"
3'-10"
10'-8"
3'-0"
49'-4"
3'-10"
4'-8"
3'-4"
19'-4"
4'-8"
3'-4"
10'-0"

5'-4"
16'-8"
8'-8"
42'-0"
16'-8"
5'-4"

## GEORGIAN HOUSE: DIMENSIONS

#12

# Elevation and Final Plot

## COMPETENCIES/LEARNING OUTCOMES

Upon successful completion of the **Elevation,** the student will have used and begun to master the following settings, operations, and drawing commands.

**Settings:** Units, Limits, Snap, Grid, Polar Tracking, Ortho, OSnap, Layers, Linetypes, Text Style, Dimension Style, **Drawing Limits**

**Operations:** Open/Exit AutoCAD; Access Menus; Enter Commands; Startup Wizard; Open; Save As and Save drawing files; Creating/Saving backup files.

**Commands:**

Line

Offset

Fillet (0)

Erase: Select, Window, Crossing Window

Undo

Zoom Window, Previous, All, Realtime

Pan

Save

Save As

Plot in Paper Space

Trim

OSnap

Fillet with Radius (R) setting

Grips

Circle

Copy and Multiple Copy

Mirror

Layer Properties

Linetype Load/Selection

Linetype Scale

Extend

Properties (Modify and Match)

Arc: SCE (Start, Center, End)

Array (Rectangular)

Break at Point

Rotate

Array (Polar)

Move

Insert

Explode (and Xplode Inherit)

Text Style, Text, DT (DText/Dynamic Text), Edit Text (ED) DText, EDit, Multiline Text

Dimension

**Hatch**

## PROCEDURE

Working alone and with classmates, incorporate information from lectures, demonstrations, and corresponding information in your text to complete this assignment.

You will be drawing a Front Elevation for the Floor Plan you have just completed. As the drawing unfolds, note the direct relationship between the width dimensions across the front of the house in Plan view and the width dimensions across the front of the Elevation. For example, you located a window 3'-4" from the corner in the Plan—that same window will show up 3'-4" from the same corner in Elevation—but will now be seen in its "height." In the office, Elevations are often "projected" directly from the Plan view, since Plan and Elevations share common dimensions. We will be drawing this Elevation "independently," but look for the relationships between Plan and Elevation just the same. You may project the width dimensions if you choose—verify with your instructor.

This assignment is accompanied by six **drawing guides (13A–13F)** that will lead you step by step in drawing an Exterior Elevation for the Georgian House. **Review this entire handout PRIOR** to beginning your drawing to understand where the individual steps are leading you and the sequence they follow. You will be asked to do a lot of **Moving** and **Copying** in this exercise—**whenever possible use your OSnaps to lock onto Base and Displacement Points when Moving and Copying.**

As always …

RELAX, EXPLORE, ASK QUESTIONS,

WORK WITH YOUR FELLOW CLASSMATES.

## ELEVATION: BLOCK OUT AND DOOR (SEE 13A)

**Open** and/or continue working on your \*\*\*GRG drawing (**FLOOR PLAN**)—your settings should already be locked in

**Make the ELEVATIONS Layer "Current"**

### BEFORE YOU BEGIN THE ELEVATION … SOME "PREP" WORK

Remember that we are assigning lineweights by Color. Referring to *Supplement 6: Quick Guide to Plotting in Paper Space*, White and Yellow lines are "wide"; Magenta is "medium thin"; and Red is "thinnest." For a small Plot, you will need to do some color work on the **Inserted** DOOR and WINDOW. They were drawn on Layer 0 and left to default to the Layer Color (Color ByLayer). Because these entities were inserted as Blocks, they must be Xploded (or Exploded) before they can be modified. The Shutter should be ok—we modified its Colors when we drew it, and AutoCAD remembers our special instructions.

Use **Zoom** for accuracy …

**Modify Properties** using the Properties Toolbar...

**Verify** that DOOR, WINDOW, and SHUTTER are on the **ELEVATIONS** Layer Xplode (Inherit) DOOR and WINDOW if you have not done so.

**Change** DOOR frame/outline to MAGENTA... a "medium thin" color

**Change** DOOR panels, flutes, and other "small" details to RED... our "thinnest" color.

**Change** WINDOW frames and "main lines" to MAGENTA.

**Change** WINDOW muntins to RED.

**You will make multiple copies of the WINDOW and SHUTTER—changing features now will save work later.**

**Ortho** ON, **OSnaps** ON

1. Lay out width and height (to bearing) of Front Elevation 49'-4" × 18'-0"

   Select an "empty" space above, below, or to the side of the Floor Plan for this drawing. Your instructor may have more specific directions for the placement and construction of this drawing. Otherwise, we will "**Move**" it into its final resting place at a later time.

2. **Offset** to build the Recessed Entry per the dimensions shown on 13A.

   Note that the FINISHED FLOOR (**Fin. Flr.**)—the horizontal reference line—is designated as **0'-0"** and that horizontal lines (heights) are **Offset** above (or below) this reference line.

   The widths shown are taken from the Floor Plan dimensions. The entry space is 19'-4" from either building corner, and is itself 10'-8" wide.

3. **Offset** Finished Grade (ground) line 1'-6" **below** Finished Floor.

   **Offset** side Porches 10'-0" left and right.

   **Extend/Trim/Fillet**—or use **Grips**—to reflect view shown in example.

4. **Place** the DOOR:

   Draw a vertical layout line from the MIDPOINT of the FINISHED FLOOR line, using the Midpoint OSnap for the beginning point. Extend this layout line up through the drawing. **Erase** the layout line after the DOOR is placed.

5. **Move** the DOOR into its final position.

   **Move** Command... select the entire DOOR

   Midpoint OSnap to **bottom** of DOOR for your Basepoint

   OSnap to the **bottom Endpoint** (or Intersection) of layout line at Entry.

Think what you want to do—you want to **Move** the MIDDLE of the door's bottom to the Endpoint of the layout line. As the Elevation is symmetrical, you really didn't need the layout line—you could have just moved the door to the Midpoint OSnap of the Finished Floor line. But since this will not always be the case, the idea of creating layout lines to locate "targets" **for Moving** (or **Copying**) is a method you should begin to practice.

This is a wonderfully accurate way to **Move** or **Copy** entities—you can OSnap to a strategic point "on" the object you want to **Move** or **Copy,** and then OSnap to a point that defines its final location. You may need to "create" that final point just for the "**Move**"—like you did here—but then those layout lines can be erased. If you try locating things by "eyeballing," you will *not* die a "thousand" deaths—you will die *ten thousand* deaths. Remember that AutoCAD's Dimensioning is absolutely accurate and unforgiving. You *must,* in turn, be accurate in your drawing. Learn to use the OSnaps.

Next we will do something similar to the WINDOW and SHUTTER. This is really neat stuff that's unfolding here.

## ELEVATION: WINDOWS AND SHUTTERS (SEE 13B)

### POSITION WINDOWS AND SHUTTERS

As you go through this next series of steps, note the similarities between the process for positioning the DOOR and WINDOW/SHUTTER assemblies. You will locate strategic points on the Elevation with layout lines—the **upper left corners** of each WINDOW, **Move** one WINDOW into place, and then, after positioning its shutters, **Multiple Copy** the assembly around the Elevation. All these processes involve the use of the OSnap in **Moving** and **Copying** the entities involved.

### POSITION ONE WINDOW

1. **Offset** the "head heights" (tops of WINDOWS) per the two horizontal lines shown.

   Note that these head heights are in **relation to Fin. Flr. (0'-0")**.

   First story window heads are at **6'-8" above 0'-0"**.

   Second story window heads are at **15'-8" above 0'-0"**. (6'-8" above the second floor level which we will assume is 9'-0")

   **Offset** the LEFT jambs (sides of WINDOWS) per the vertical lines shown.

   Note the direct relationship between these dimensions and the dimensions used to locate the left jambs when you laid out the Floor Plan. Remember that these are the "same" windows—the same building features—you see in Plan View. You are just drawing them from a different viewpoint, in Elevation instead of in Plan. Also note that you are *not* drawing the actual WINDOWS. Rather you are giving yourself OSnap "reference points" with layout lines, which will allow you to place the WINDOWS accurately using the OSnap.

2. **Move** one WINDOW into place.

   **Move … select** (OSnap) upper left "inside" corner of the frame as Base Point.

   **Select** (OSnap) **Intersection** of upper left layout lines as the Displacement Point.

   As was the case for the DOOR, think in terms of "*from here*"—"*to there*," using OSnaps as your guide. You may need to do a lot of "Zooming IN and OUT" during this process. Remember that the **Zoom** command when accessed from the Toolbar is a "transparent" command. This means that you can perform "Zooms" while in the middle of a **Move** or **Copy** or **Multiple Copy** command. Take advantage of this feature.

   Using OSnaps for Base and Displacement points is absolutely accurate, providing the reference points you created are accurate.

### POSITION SHUTTERS

3. Place a SHUTTER at each side of the WINDOW.

   Using OSnaps for the Base and Displacement Points …

   First, **Copy (not Move)** the SHUTTER to one side of the WINDOW.

   OSnap "from" an upper SHUTTER corner to the corresponding jamb/head Intersection at the WINDOW.

   Then **MOVE** the original SHUTTER to the opposite side of the WINDOW,

   following the same procedure.

   Your drawing should now look like the example at the upper right of 13B.

4. OK—this is going to be really cool!!

   **MULTIPLE COPY** the WINDOW/SHUTTER assembly you have created.

   **Copy** … capture the entire WINDOW/SHUTTERS assembly in a Window.

   Use the jamb/head Intersection as your Base Point, **Zooming** IN to capture it **precisely with your OSnap**

When prompted, select "**M**" for **Multiple Copy,** Enter  (the Multiple option defaults for release 2005)

**OSnap** to each of the seven remaining Jamb/head Intersections you created with your layout lines.

When Moving and/or Copying, if you want to "see" the WINDOW/SHUTTER assembly tracking with your cursor, Ortho must be OFF (unless special "system variable settings are selected). But even if Ortho is ON, if you locate an OSnap, entities will "jump" to the desired location.

If you go astray you can always cancel and retry, or **Undo** and retry.

**ERASE** the layout lines when you are sure that WINDOWS/SHUTTERS are properly located.

Your drawing should now look like the example at the bottom of 13B (with the reference lines erased).

## ELEVATION: ROOF, FASCIA, AND CHIMNEY (SEE 13C)

As you proceed through this series of exercises, note the power and convenience of the **Offset** command, and especially how you can use it in conjunction with **Extend, Trim, Fillet,** and/or **Grips** to construct and locate items quickly and *accurately*. Think of these offsets as measuring lines or layout lines that will be shaped, modified, or later erased as the final elements take form.

Many of the following operations will need to be repeated on "each side" (EA.SD.) of the Elevation. You may simply "repeat" the operations or draw "one side" and **Mirror** to the other, OSnapping a Mirror Line with Ortho ON from the Midpoint of the facade.

**Look at the "finished" FASCIA and TRIM piece before you begin this section (number 4).** This will help you understand the process and your goal. You will see that the TRIM piece extends 2" from the face of the Fascia board, and that the roof line extends from the top of the TRIM piece to the Ridge line. Don't confuse the 2 × 2 "trim piece" on the FASCIA with the "**Trim**" command when reading these instructions.

The roof structure is a Mansard "style" roof (not a true Mansard)—it slopes steeply at the outside, but forms a Parapet for a flat roof at its inside.

## DRAW ROOF WITH FASCIA AND TRIM

1. **Offset** for roof's Ridge height of 3'-0" above the top of walls —

    this would be 21'-0" above Fin.Flr. (0'-0") ... right?

    **Offset** each side 2'-0" for the Overhang (Eave).

2. **Extend** (or **Fillet**) the top of the wall to form corners between the wall and overhang Offsets. This horizontal line will now represent the "top" of the Fascia's 2 × 2 Trim piece.

3. **Offset below** the line as shown 2" for the trim's height (it's 2 × 2) and another 6" for the bottom of the Fascia board—or 8" total from the top of the Trim piece.

4. **Trim** the face of the overhang as shown in the drawing top right of 13C.

    **Zoom** for detail work at the Fascia, Trim Piece, and Roof.

    **OSnap** an angled line from the top of the Trim piece to the Ridge line.

    **Offset** the "face of the Fascia" 2" from the face of the Trim piece.

    **Trim** the overhang detail as shown.

    Repeat the entire process at the opposite overhang.

## DRAW CHIMNEY

The dimensions given in this exercise are based on the examples shown in the previous assignments. If you have located your FIREPLACE differently, or have added a FIREPLACE at other places in the house, then adjust your Elevation accordingly.

5. **Offset** from the left corner of the building 12" for the outside wall of CHIMNEY.

    **Offset** 24" from the Ridge for the top of the CHIMNEY stack.

**Fillet** these two lines to form a corner.

**Offset** from the outside wall of CHIMNEY 2'-8" for the Chimney's width.

Refer to the Floor Plan if you're unsure what's happening here. The "front face" of the FIREPLACE/CHIMNEY is "inside" the building—we don't see it until it pokes through the roof.

**Trim** as shown at the Roof and bottom where it "blocks" part of the Porch from view.

Using **Offset, Trim, Fillet, Grips** ... construct the Flue and Grout Cap as shown.

## ELEVATION: PORCHES AND FRONT STEPS (SEE 13D)

### SIDE PORCHES

1. Follow the procedure shown on 13D to draw the Pillars at each Side Porch.

    The Pillars at the Side Porches, as shown on your Floor Plan, are 32" square.

    Draw them 32" wide × 32" high in their side Elevation.

    **Offset, Fillet, Trim, Grips** will serve you well. Note the similarities between strategies for drawing the Pillars and the Chimney, Fascia, and Elevation.

### FRONT PORCH and STEPS

The Front Steps are visible in this Elevation view, between flanking Pillars.

1. Draw Pillars at the Front Porch 24" wide × 36" high.

2. **Offset** the front step Risers 6" each.

3. **Trim** the Riser lengths accordingly.

Your ELEVATION should now look like the drawing at the bottom of 13D.

## ELEVATION: ALTERNATES (SEE 13E)

Shown on 13E are three possible alternates for "Georgian Elevations." You may make your own variations as time permits, but work from the basic elevation you have constructed. Verify your intentions with your instructor.

Make good use of the **Copy, Multiple Copy, Mirror,** and/or **Array (Rectangular)** as you construct these options. Use your own judgment for unspecified dimensions of details—look to the examples for general guides.

For the **BALUSTER, lay in the Rail height and thickness at the right side Porch first,** then construct a layout line from the bottom of the Rail to the Porch Floor. Use this line as a "true height" axis around which the Baluster form may be constructed. The Baluster shown was constructed on a vertical axis by over-lapping a series of **Circles** and **Ellipses,** connecting with and to straight lines, and then **Trimming.** The **Tangent OSnap** may also prove useful. Your Baluster does not have to appear exactly like the example.

When locating items on your drawing, remember to construct "reference points" to which you can OSnap for precise locations. **Erase** these construction lines when you are through with them.

### HATCHING

**Read all this information** *before* **Hatching.**

**Always SAVE** *before* **hatching.**

Create a new Layer named HATCH. Color: RED (we will want the Hatch to Plot "thin"). **Make the Hatch Layer Current.**

1. **Select Hatch** command in the **Draw** Pulldown menu or through the Hatch icon in the Draw Toolbar ... opening the **Boundary Hatch and Fill** dialog box.

2. **Selecting** a **Hatch Pattern:**

    AutoCAD provides a wide variety of hatching patterns at a variety of scales. By selecting the "blank" button to the right of the **Pattern** flyout arrow, the Patterns will be graphically displayed in the **Hatch Pattern Palette** dialog box. Architectural formats typically use **ANSI31 for Concrete Masonry Units (CMU)** at walls in Plan View. For other options typically used in Elevations (and Plans), select the **Other Predefined** tab on the Hatch Pattern Palette. Your instructor may have suggestions for Patterns that will serve you best.

3. **Setting a Hatch Scale** ... **Begin with a Scale of 24**

The Pattern options default to a variety of sizes or scales and until you become familiar with a few of the standard Patterns, it's wise to **begin with a Scale of 24.** This will bring in the Pattern 24 times larger than the default size. Through the **Preview** option, you can then view the Pattern and readjust the Scale—smaller or larger—as desired, **Previewing** as you go. For some—but not all—the Patterns, starting out with the default Scale of 1 may appear to "fill the hatched area solid," when in reality AutoCAD is placing the pattern very very small on our very very large drawings. Conversely, with the safe starting Scale of 24, the Pattern may come in so large that the selected region might appear "empty." In those cases, keep readjusting Scale until you reach desired size.

4. **Selecting** the **Hatch Area:**

Either the **Pick Points** or **Select Objects** tabs will allow you to select the Hatched area. **Pick Points** allows you to simply select the bounded region by picking any random internal point within the boundary. **Select Objects** will force you to individually select the boundary lines. Typically the **Pick Points** option works best.

**Select** the **Pick Points** button at the upper right of the dialog box.

**Select** (Left Click) an area or region to Hatch.

Three things to consider here ... First of all, you need to have the **entire region** you wish to **Hatch** visible on the screen before entering the **Hatch** command—AutoCAD will not let you Zoom Out or **Pan** to capture areas not already displayed. If you try to **Hatch** a region that bleeds off the screen, you will get a message telling you **VALID HATCH BOUNDARY NOT FOUND.** You'll need to cancel the **Hatch** command and readjust your screen.

Second, you need to have **"closed" boundaries**—if you have a "leak" somewhere (typically at a corner), you will either receive the invalid boundary message or you will hatch additional, unintended areas. You can create temporary boundary lines, breaking the **Hatch Area** into smaller parts to isolate the leaky trouble causer. Also the **Hatch** command can be persnickety, sometimes not liking boundaries with hidden, miscreant line segments below, blocks touching them, and the like. If you're running into problems, best to ask your instructor for some expert advice.

Third, **you may select more than one region at a time**—provided they are on your screen when you enter the command. You might be best served to select a small area first, **Preview** and adjust to your satisfaction, then Pick more areas before "Okaying" your way home. The drawback to picking multiple areas is that the **Hatch patterns come into your drawing as a single entity and cannot be edited separately.** This means that you can't **Trim** or **Extend** Hatch Patterns in our typical fashion should you need to make such adjustments down the road, nor can you **Erase or Move just a portion** of the Hatch Pattern. So while it's efficient to Pick several areas for Hatching in one operation, it's probably not wise to Hatch the entire drawing, especially if it's a complex drawing, in one giant swoop.

**Right Click** to return to the dialog box for Preview options/readjustment of Scale

**OK** to accept the **Hatch**

---

## HATCHING – IT'S IN, NOW WHAT DO I DO WITH IT?

Hatch Patterns can be modified through the **Modify** Pulldown/**Objects** or through the **Properties** Dialog box. The **List** command will give you information about a Hatch in place. A shortcut here is to "double click" on the Hatch pattern and the **Edit Hatch** dialog box appears.

You can **Erase, Mirror, Match Properties** (paint brush), etc. **Hatch** patterns, as well as assign them to different Layers or Colors. You generally want the Hatch to Plot "thin," which makes RED a good choice for us.

Some assign certain Hatch patterns a Color that is "screened" when plotted to produce a lighter shade of gray on the final plot, contrasting with the black, monochrome line work. In these cases, you must send the Hatch "to the back" (**Tools/Display Order/Send to Back**) so that black lines plot "over the top" of the gray lines, and not vice-versa.

## PLAN AND ELEVATION: FINAL FORMATS, NOTES, PLOT (SEE 13F)

We're setting out to play with our **Drawing Limits** here, and intend to Plot in Portrait on 11 × 17 paper. Verify these plotting parameters with your instructor before proceeding.

Think of your existing paper **Area** (**Drawing Limits**) in terms of $(x,y)$ coordinates: the bottom left corner being (0,0) and the current upper right corner being (144',96')—144' *across* to the right × 96' *up*. Your paper is now lying in what we call "**Landscape**" orientation with the long side horizontal. You will be asked to make a final Plot of the Georgian House in "**Portrait**" style, with the long side vertical, which means to position things correctly the **Drawing Limits** must be reset for a piece of paper 96' WIDE (across) × 144' LONG (up and down)... or the opposite of what you now have.

Because we are plotting in **Paper Space**, it's not really necessary that we play with the **Drawing Limits** like this—we can manipulate views in **Paper Space** through Viewport management. But as was the case with previous assignments, we're after experience with a variety of AutoCAD commands and options. Look at this portion of the exercise as an opportunity to work with and readjust **Drawing Limits**.

1. **Reset** your **Drawing Limits**

   **Format** Pulldown/**Drawing Limits**

   > default to the **lower left** corner at (0,0); **Enter**

   > **96', 144'** (you *must* type the comma) for the **upper right** corner; **Enter**

2. Turn Grid ON (if it's OFF) and **Zoom All** to display the area bounded by the redefined drawing LIMITS.

3. **Make sure that ALL Layers are turned ON** and then **Move** the Plan and Elevation to the regions shown on 13F. Align the Elevation with the walls of the Floor Plan.

4. Make your **Text Layer** "Current"; **DTEXT (DT)** command for Titles, height 10"

   Place drawing titles—FLOOR PLAN and ELEVATION.

   Try underlining with a **Polyline** (**Draw** Pulldown or Toolbar) Width = 3".

   ...A new command that will be fun to figure out.

   > Pick start point, Enter **W** for Width, Enter Width for each end of line, Pick end.

   > **Pedit** will allow you to Edit this line.

5. **DTEXT** ... height 6" ... Place Project Title (Georgian House), your name, date.

6. **Plot** your finished drawing in **Paper Space**. 11 × 17 (tabloid), Portrait, $^1/_8$" = 1'-0", Lineweights for Small Plot. If an 11 × 17 Plotter is not available, try Legal $8^1/_2$ × 14 (Legal) and "Scale to Fit." Remember to deselect the box on Format/Linetype, and to move Viewport to the VIEWPORT Layer. Verify the size and scale of the final Plot with your instructor ... that will influence your Lineweight settings.

### THERE!

**Take a minute to look back at what you've done—how far you've come since Assignment 1.**

**That PANEL DOOR doesn't look so tough now, does it!?**

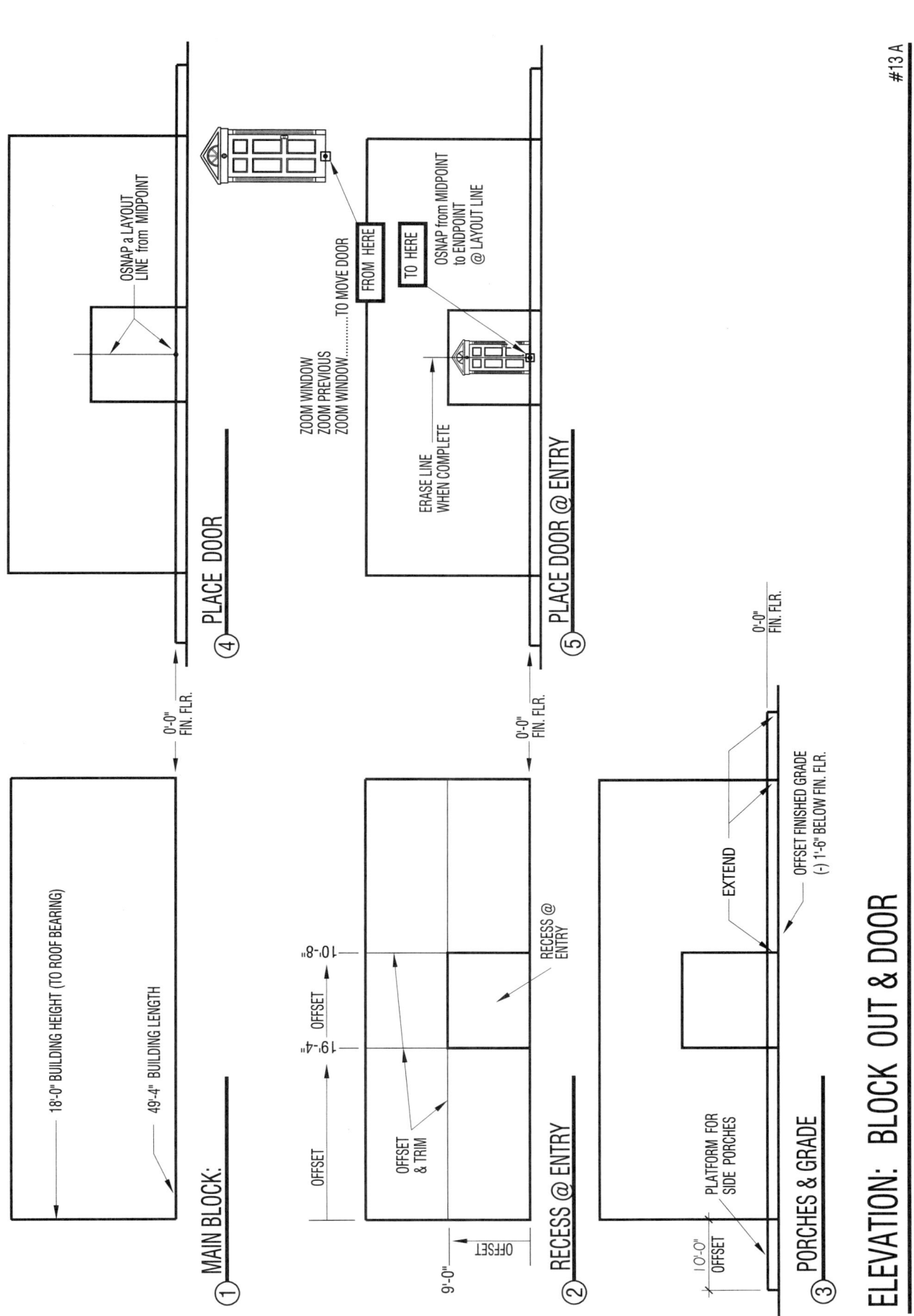

ELEVATION: BLOCK OUT & DOOR

① MAIN BLOCK:

18'-0" BUILDING HEIGHT (TO ROOF BEARING)

49'-4" BUILDING LENGTH

② RECESS @ ENTRY

OFFSET

OFFSET & TRIM

9'-0" OFFSET

10'-8"

19'-4"

RECESS @ ENTRY

③ PORCHES & GRADE

PLATFORM FOR SIDE PORCHES

10'-0" OFFSET

EXTEND

OFFSET FINISHED GRADE (-) 1'-6" BELOW FIN. FLR.

0'-0" FIN. FLR.

④ PLACE DOOR

OSNAP a LAYOUT LINE from MIDPOINT

0'-0" FIN. FLR.

ZOOM WINDOW
ZOOM PREVIOUS
ZOOM WINDOW........TO MOVE DOOR

FROM HERE

TO HERE

OSNAP from MIDPOINT to ENDPOINT @ LAYOUT LINE

ERASE LINE WHEN COMPLETE

⑤ PLACE DOOR @ ENTRY

0'-0" FIN. FLR.

#13 A

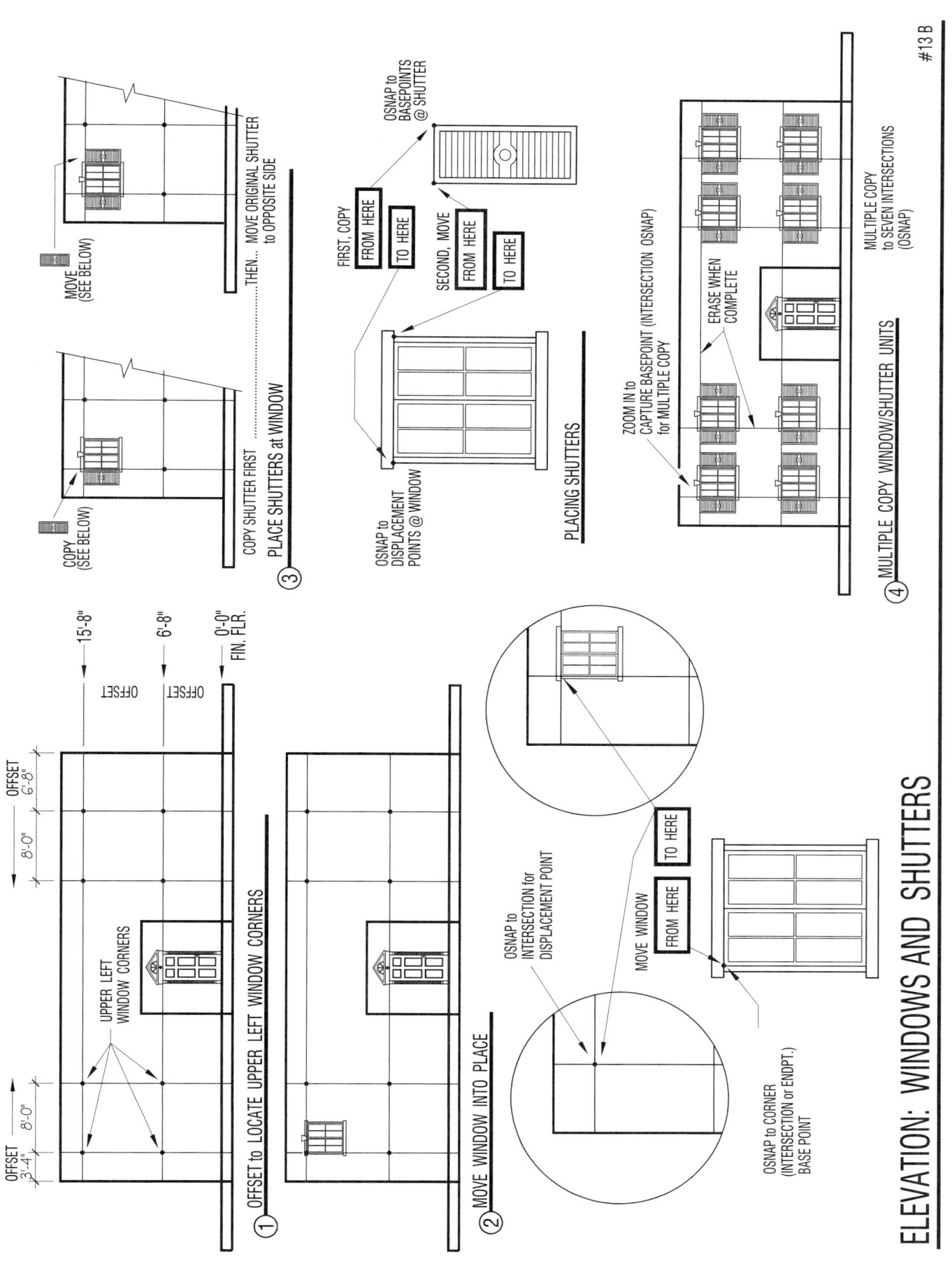

OSNAP to
BASEPOINTS
@ SHUTTER

FIRST, COPY
FROM HERE
TO HERE

SECOND, MOVE
FROM HERE
TO HERE

OSNAP to
DISPLACEMENT
POINTS @ WINDOW

PLACING SHUTTERS

MOVE
(SEE BELOW)

......THEN... MOVE ORIGINAL SHUTTER
to OPPOSITE SIDE

COPY
(SEE BELOW)

COPY SHUTTER FIRST

③ PLACE SHUTTERS at WINDOW

ERASE WHEN
COMPLETE

ZOOM IN to
CAPTURE BASEPOINT (INTERSECTION OSNAP)
for MULTIPLE COPY

MULTIPLE COPY
to SEVEN INTERSECTIONS
(OSNAP)

④ MULTIPLE COPY WINDOW/SHUTTER UNITS

15'-8"

6'-8"

0'-0"
FIN. FLR.

OFFSET

OFFSET

OFFSET
6'-8"

8'-0"

OFFSET
3'-4"

8'-0"

UPPER LEFT
WINDOW CORNERS

① OFFSET to LOCATE UPPER LEFT WINDOW CORNERS

OSNAP to
INTERSECTION for
DISPLACEMENT POINT

MOVE WINDOW
FROM HERE
TO HERE

OSNAP to CORNER
(INTERSECTION or ENDPT.)
BASE POINT

② MOVE WINDOW INTO PLACE

ELEVATION: WINDOWS AND SHUTTERS

#13 B

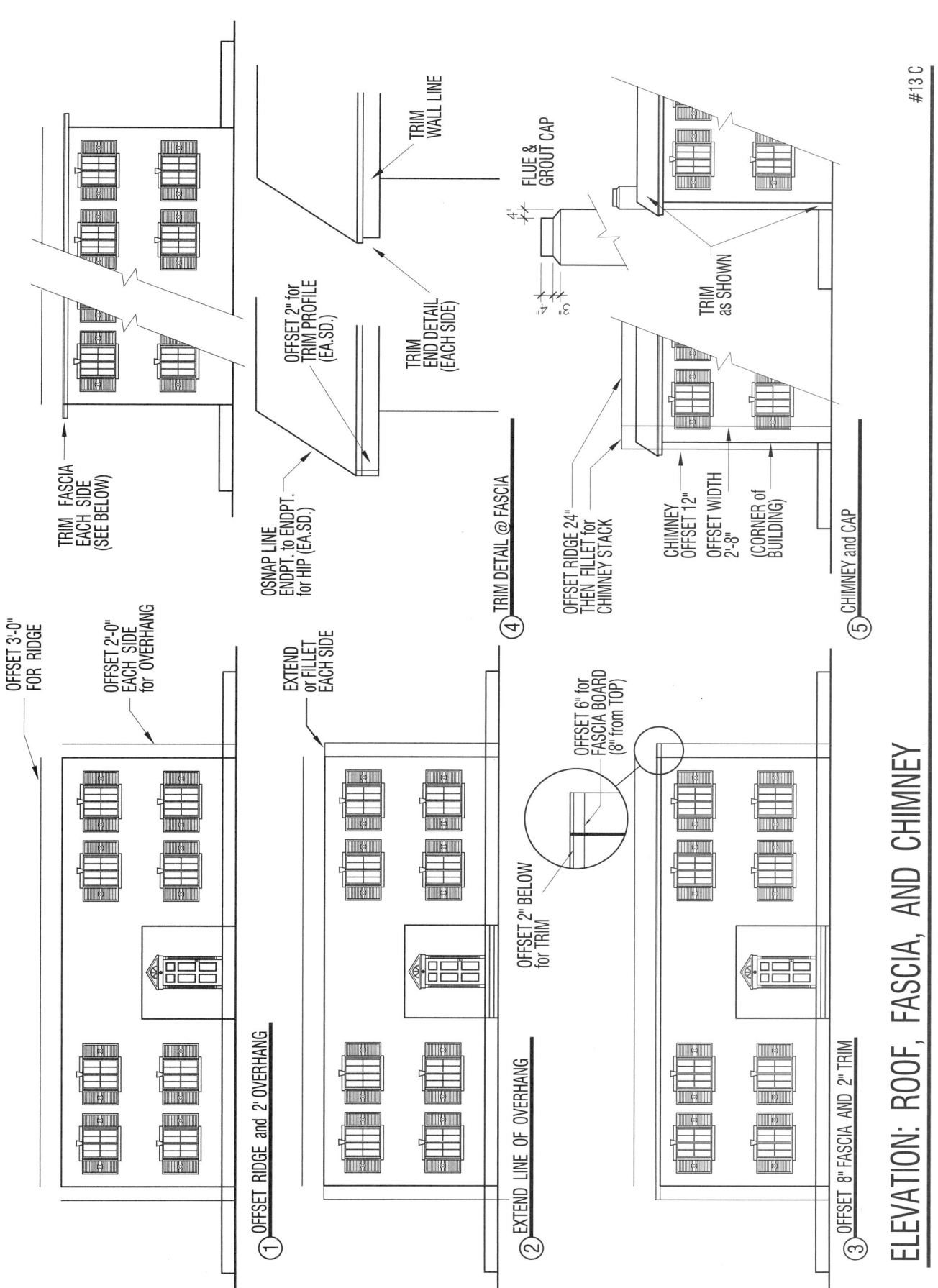

OFFSET 3'-0"
FOR RIDGE

OFFSET 2'-0"
EACH SIDE
for OVERHANG

EXTEND
or FILLET
EACH SIDE

OFFSET 2" BELOW
for TRIM

OFFSET 6" for
FASCIA BOARD
(8" from TOP)

① OFFSET RIDGE and 2' OVERHANG

② EXTEND LINE OF OVERHANG

③ OFFSET 8" FASCIA AND 2" TRIM

TRIM  FASCIA
EACH SIDE
(SEE BELOW)

OFFSET 2" for
TRIM PROFILE
(EA.SD.)

TRIM
WALL LINE

TRIM
END DETAIL
(EACH SIDE)

OSNAP LINE
ENDPT. to ENDPT.
for HIP (EA.SD.)

④ TRIM DETAIL @ FASCIA

FLUE &
GROUT CAP

4"

3"

TRIM
as SHOWN

OFFSET RIDGE 24"
THEN FILLET for
CHIMNEY STACK

CHIMNEY
OFFSET 12"

OFFSET WIDTH
2'-8"

(CORNER of
BUILDING)

⑤ CHIMNEY and CAP

ELEVATION:  ROOF,  FASCIA,  AND  CHIMNEY

#13 C

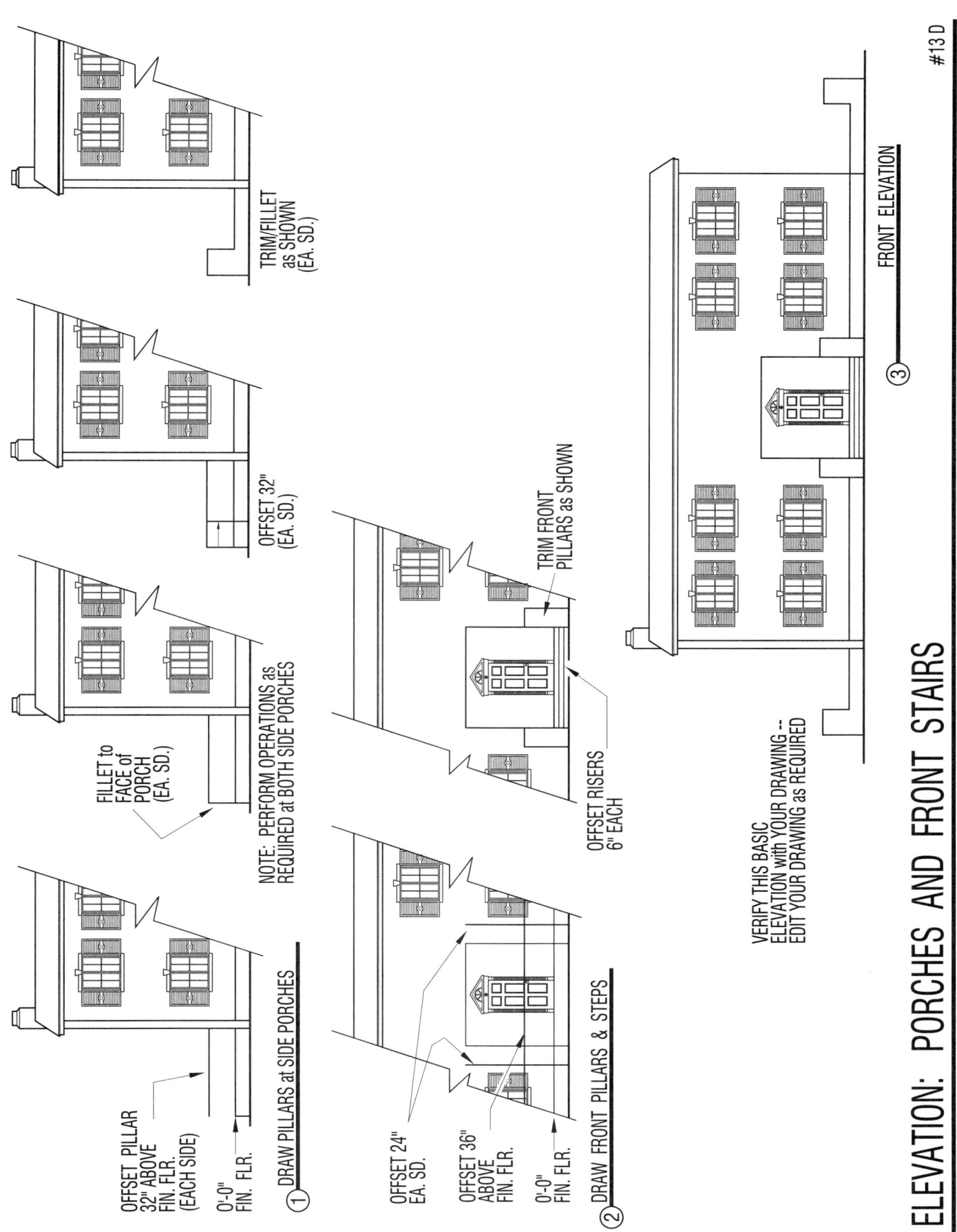

OFFSET PILLAR 32" ABOVE FIN. FLR. (EACH SIDE)

0'-0" FIN. FLR.

FILLET to FACE of PORCH (EA. SD.)

NOTE: PERFORM OPERATIONS as REQUIRED at BOTH SIDE PORCHES

OFFSET 32" (EA. SD.)

TRIM/FILLET as SHOWN (EA. SD.)

① DRAW PILLARS at SIDE PORCHES

OFFSET 24" EA. SD.

OFFSET 36" ABOVE FIN. FLR.

0'-0" FIN. FLR.

② DRAW FRONT PILLARS & STEPS

TRIM FRONT PILLARS as SHOWN

OFFSET RISERS 6" EACH

VERIFY THIS BASIC ELEVATION with YOUR DRAWING -- EDIT YOUR DRAWING as REQUIRED

③

FRONT ELEVATION

ELEVATION: PORCHES AND FRONT STAIRS

#13 D

13'-0"

9'-0"

18" DIA.
COLUMNS
4x22 BASE
3x22 CAPITAL

0'-0"

8" FASCIA
2" TRIM

DESIGN ONE BALUSTER
and MULTIPLE COPY
or ARRAY

2" RAIL + 30"

HATCH:
BEGIN with SCALE FACTOR of 24 --
ADJUST UP or DOWN from THERE (PREVIEW),
ALWAYS SAVE BEFORE ATTEMPTING ANY HATCHING!!!!!!

COLUMNS, TRIANGULAR PEDIMENT, and RAILS

48" DIA. LUNETTE

2"
6"
8"

13'-0"

9'-0"

24" DIA.

RAILED ENTRY

2x4

CONSTRUCT
AROUND
AN AXIS

2x6

POSSIBLE
BALUSTER

ROUND WINDOW
24" &/or 48" DIA.
POLAR ARRAY SPOKES
COPY/SCALE COMMAND

CONC. SPHERE
12" DIAMETER
2x8 BASE

ROUND WINDOWS, LINTEL, CONCRETE SPHERES w/ BASE

# ELEVATIONS: THREE ALTERNATES

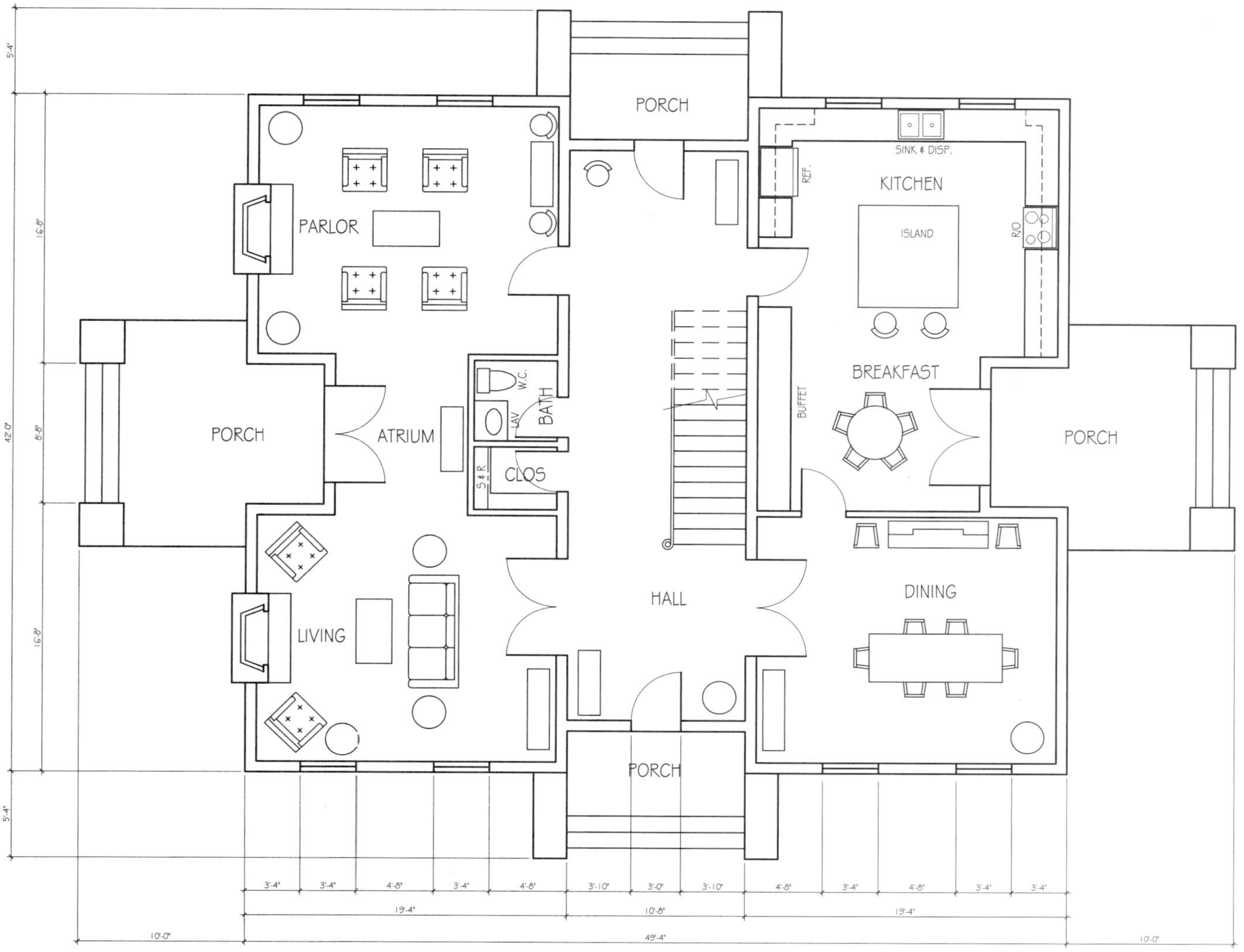
PORCH
PARLOR
KITCHEN
SINK & DISP.
REF.
ISLAND
R/O
BREAKFAST
BUFFET
W.C.
BATH
LAV
PORCH
ATRIUM
S&R
CLOS
PORCH
DINING
HALL
LIVING
PORCH
5'-4"
16'-8"
8'-8"
42'-0"
16'-8"
5'-4"
3'-4"
3'-4"
4'-8"
3'-4"
4'-8"
3'-10"
3'-0"
3'-10"
4'-8"
3'-4"
4'-8"
3'-4"
3'-4"
19'-4"
10'-8"
19'-4"
10'-0"
49'-4"
10'-0"

FLOOR PLAN

FRONT ELEVATION

GEORGIAN HOUSE

DRAWN BY: (YOUR NAME)
DATE:

(13F)

GROUP

2

BROWNSTONE

# Upper Level Plan and Lower Level Plan

The Brownstone project is a two-story masonry structure built in the Victorian/Gothic Revival style of Frank Furness, an American architect working in the later part of the nineteenth century. Originally designed as an office building, it has since been remodeled to serve as an Art Gallery on the Lower Level and a residence for the gallery's owner at the Upper Level. This project has been carefully shaped to work as a springboard into more complex work, providing you an opportunity to apply lessons first introduced through the Georgian series.

Instructions will lead your way, but they are much briefer than those for the Georgian Unit, as you now will begin to rely on your developing skills. Refer to previous assignments as needed and ask questions of your instructor for clarifications or "reminders." Follow instructions and steps as shown, working on your own and with your fellow students.

**SAVE BROWNSTONE WORK TO YOUR DESIGNATED DRIVE**
**SAVE BACKUPS TO A DISK**

**Complete Assignments 14 and 15 in the same drawing file.**

**Setup:**

**Use a Wizard** (**Quick** setup) to setup a new drawing.

**Units: Architectural**

Pick **Next**

**Area: 144' by 96'** (limits for ¼" scale on a 36" × 24" paper—see **Supplement 2: Drawing Area…**)

Pick **Finish**

**Zoom All** to capture your full screen.

**Save As** to name this drawing—your three initials BRNSTN. Example: MRB BRNSTN.

See **Supplement 5: Floor Plan Linetypes and Layers** and use materials (as needed) from the Georgian House; set the following from the *Format pulldown*:

**Load Linetypes:**

Hidden, Hidden2, Hiddenx2, Center, Center2, Centerx2, Phantom

Set **Global Linetype Scale** to 24 (half the ¼" scale factor of 48—see **Supplement 2:…**)

Create **New Layers**—follow **Supplement 5: Floor Plan Linetypes and Layers**

Set/Name New **Text Style**: "Architectural 48"— per Georgian House, Assignment 11

Set **Dimension Style**: per Georgian House, Assignment 12

Set Running **OSnaps** (Tools Pulldown/Drafting Settings)

    Endpoint, Midpoint, Center, Intersection, and Perpendicular.

Verify **Fillet Radius = 0"**

OK...go to the Tools/Drafting Settings pulldown and Set a Grid to 48". We won't be using the grid for much at this point, but it gives you a look at the boundaries of your paper. When you get the outline of your building laid out, you may need to use the **Move** command to reposition it on the paper. No need to keep the Grid ON except for an occasional spot check on where your drawing is located with respect to your paper.

**Procedure:**

Refer to drawings/instructions for Upper and Lower Level Plans, including the $17 \times 11$ drawing at the back of the text. You will want to work from the big to the small; from the general to the particular. Lay out the perimeter first, then **Offset** the exterior walls, then furring, then the interior partitions, with no attention to the location of doors, windows, stairs, cabinets, etc. Once the walls have been positioned, then door and window openings (but not the doors and windows themselves) are located, followed by cabinets, stairs, doors, windows, plumbing fixtures, and smaller details. Hatching, Dimensions, and Text are last.

Assign entities to proper Layers in batches as you work. Be sure that walls are resting on the correct Layers before you begin other more detailed work.

Follow this general process whenever you layout a Floor Plan—it's the most efficient and minimizes "erase" time.

**Complete** the **Upper Level Plan** first (including Dimensions and Text), then **Copy** the plan directly below—with Ortho ON. You then can erase and modify the carcass to begin building the **Lower Level Plan** on the same drawing.

**Remember to SAVE your work at frequent intervals**

## PLOTTING THE FLOOR PLANS
### INSERT TITLE BLOCK

To finish off your work, insert the Title Block provided in the Blocks folder (from your instructor through the text's website) or another of your instructor's choice. The provided Title Block was created for a $36 \times 24$ sheet of paper. **Insert** it into **Paper Space** (not Model Space) with the **Insertion Point** specified for (0,0) and **Scale Factors** of 1. You may have to **Move** it a small bit to adjust for different Plotter margins. Make sure that it doesn't end up on the Viewport Layer after Insert—that Layer won't Plot, remember?

To Insert the Title Block into smaller formats, still in **Paper Space**, in the Insert dialog box, adjust the **x** Scale Factor by a ratio of the Plotting width over 36, and the **y** Scale Factor by the Plotting length over 24. For example, to Insert into a $17 \times 11$ sheet, the **x** ratio would be 17/36 and the **y** ration would be 11/24. This also reduces the Title Block's Text height, but in most cases that's OK for a check Plot.

**Placing Text in Paper Space** is exactly like placing Text in Model Space, with one important exception. Text heights in Paper Space aren't artificially adjusted. If you need Text $1/4$" high on the final Plot, then enter a height of $1/4$" when placing the Text. Complete the title Block information by placing Text in Paper Space. Match the Text heights already in place.

> Your instructor may introduce you to **Attributes** in connection with the Title Block information... or this may be waiting in a later, more advanced class.

## PLOTTING

You know the drill by now for plotting. If your Plotter(s) supports the option, try $36 \times 24$ Plots of each Plan, scaled to $1/4$" = 1'-0". Use LARGE Lineweight settings (see *Supplement 6: Quick Guide...*).

Pay special attention to Lineweights and exercise patience and perseverance in creating a Plot of which you will be proud. Examining the Plot for Lineweights is actually a crucial part of your learning. There's no better way to develop your "on screen" Color/Lineweight intuition than to develop a picture in your mind of the final result.

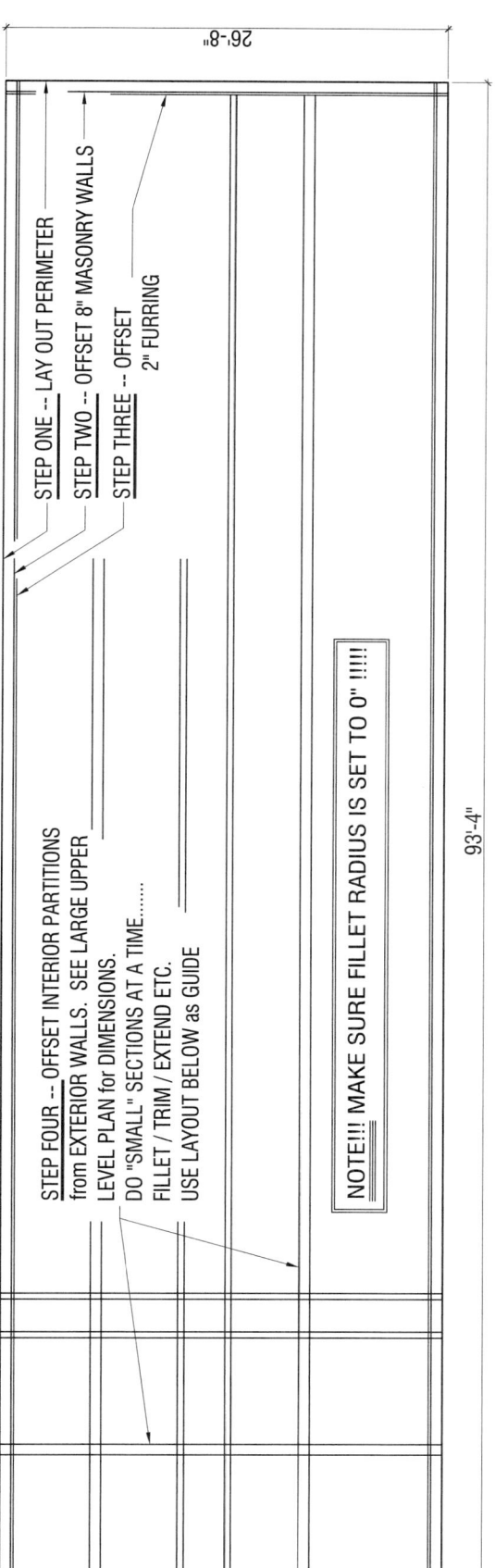

26'-8"

93'-4"

STEP ONE -- LAY OUT PERIMETER

STEP TWO -- OFFSET 8" MASONRY WALLS

STEP THREE -- OFFSET 2" FURRING

STEP FOUR -- OFFSET INTERIOR PARTITIONS
from EXTERIOR WALLS.  SEE LARGE UPPER
LEVEL PLAN for DIMENSIONS.
DO "SMALL" SECTIONS AT A TIME......
FILLET / TRIM / EXTEND ETC.
USE LAYOUT BELOW as GUIDE

NOTE!!! MAKE SURE FILLET RADIUS IS SET TO 0" !!!!!

8" MASONRY EXTERIOR WALLS (CROSS HATCHED WALLS)
2" FURRING-- 1x2"s w/ 1/2" GYPSUM BOARD
4.5" INTERIOR PARTITIONS -- 2x4's w/1/2" GYP. BD. EA.SD.
6.5" PUMBING WALLS-- 2x6's w/ 1/2" GYP. EA.SD.

LAYOUT PERIMETER WALLS w/ORTHO "ON" -- DRAG and TYPE
OFFSET 8" to the INTERIOR SIDE for MASONRY.
OFFSET 2" to the INTERIOR SIDE for THICKNESS of FURRING.
LAYOUT INTERIOR PARTITIONS USING the OFFSET COMMAND.

PLUMBING WALLS

8" WIDE

8" WIDE

LOCATE CENTERPOINT,
DRAW CIRCLE --
SNAP to WALL for RADIUS.

EXPOSED MASONRY WALL --
NO FURRING.

DO a "ROUGH" TRIM at WALLS -- WAIT UNTIL DOORS and WINDOWS
are PLACED BEFORE CLEANING UP WALL INTERSECTIONS.

DO NOT PLACE DOORS and WINDOWS AT THIS POINT --
JUST BLOCK OUT BASIC PARTITION LOCATIONS.

MODIFY PROPERTIES -- MOVE INTERIOR PARTITIONS to "WALLS_INT" LAYER.

BROWNSTONE:  UPPER LEVEL BLOCK OUT

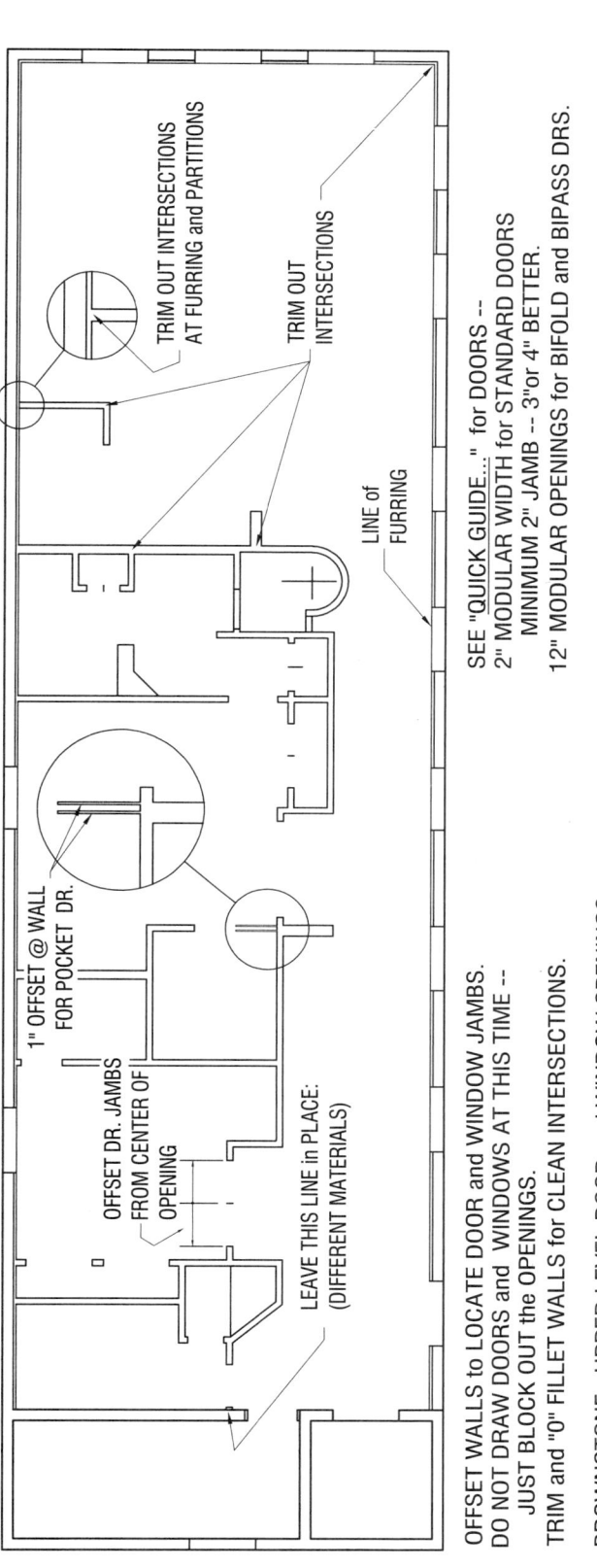

TRIM OUT INTERSECTIONS AT FURRING and PARTITIONS

TRIM OUT INTERSECTIONS

LINE of FURRING

1" OFFSET @ WALL FOR POCKET DR.

OFFSET DR. JAMBS FROM CENTER OF OPENING

LEAVE THIS LINE in PLACE: (DIFFERENT MATERIALS)

SEE "QUICK GUIDE..." for DOORS --
2" MODULAR WIDTH for STANDARD DOORS
MINIMUM 2" JAMB -- 3" or 4" BETTER.
12" MODULAR OPENINGS for BIFOLD and BIPASS DRS.

OFFSET WALLS to LOCATE DOOR and WINDOW JAMBS.
DO NOT DRAW DOORS and WINDOWS AT THIS TIME --
JUST BLOCK OUT the OPENINGS.
TRIM and "0" FILLET WALLS for CLEAN INTERSECTIONS.

BROWNSTONE: UPPER LEVEL DOOR and WINDOW OPENINGS

16" COUNTER

$4\frac{1}{2}$" REVEAL

24" COUNTER

12'-3"

CENTERLINE of STAIR

4'-2"

8'-0"

LAY IN STAIR LANDINGS and CENTERLINE
ALLOCATE ENTITIES to APPROPRIATE LAYERS -- ALWAYS -- AS YOU DRAW
USE "CURRENT" LAYER, CHANGE PROPERTIES,
& MATCH PROPERTIES COMMANDS

OFFSET WALLS to LOCATE CABINETS/SHELVES/RODS/STAIRS.
USE GRIPS to EXTEND as REQUIRED -- TRIM as REQUIRED.
WALL CABS are GREEN, HIDDEN LINES.
RODS are GREEN, CENTER LINES.

BROWNSTONE: UPPER LEVEL BUILT-INS

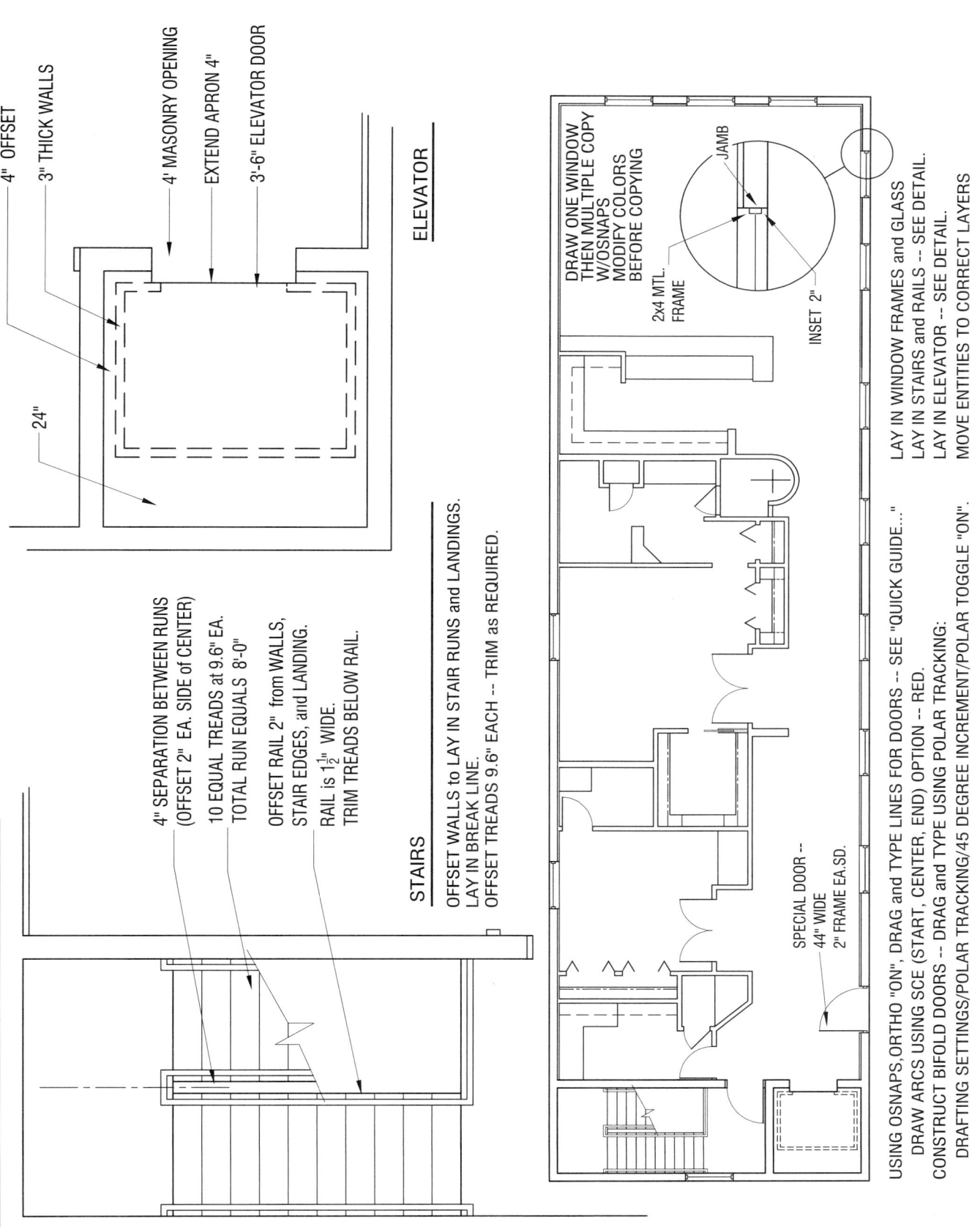

4" OFFSET

3" THICK WALLS

4' MASONRY OPENING

EXTEND APRON 4"

3'-6" ELEVATOR DOOR

24"

ELEVATOR

4" SEPARATION BETWEEN RUNS
(OFFSET 2" EA. SIDE of CENTER)

10 EQUAL TREADS at 9.6" EA.
TOTAL RUN EQUALS 8'-0"

OFFSET RAIL 2" from WALLS,
STAIR EDGES, and LANDING.

RAIL is $1\frac{1}{2}$" WIDE.
TRIM TREADS BELOW RAIL.

STAIRS

OFFSET WALLS to LAY IN STAIR RUNS and LANDINGS.
LAY IN BREAK LINE.
OFFSET TREADS 9.6" EACH -- TRIM as REQUIRED.

DRAW ONE WINDOW
THEN MULTIPLE COPY
W/OSNAPS
MODIFY COLORS
BEFORE COPYING

JAMB

2x4 MTL.
FRAME

INSET 2"

SPECIAL DOOR --
44" WIDE
2" FRAME EA.SD.

USING OSNAPS, ORTHO "ON", DRAG and TYPE LINES FOR DOORS -- SEE "QUICK GUIDE..."
DRAW ARCS USING SCE (START, CENTER, END) OPTION -- RED.
CONSTRUCT BIFOLD DOORS -- DRAG and TYPE USING POLAR TRACKING:
DRAFTING SETTINGS/POLAR TRACKING/45 DEGREE INCREMENT/POLAR TOGGLE "ON".

LAY IN WINDOW FRAMES and GLASS
LAY IN STAIRS and RAILS -- SEE DETAIL.
LAY IN ELEVATOR -- SEE DETAIL.
MOVE ENTITIES TO CORRECT LAYERS

BROWNSTONE: DOORS, WINDOWS, STAIRWAY, ELEVATOR

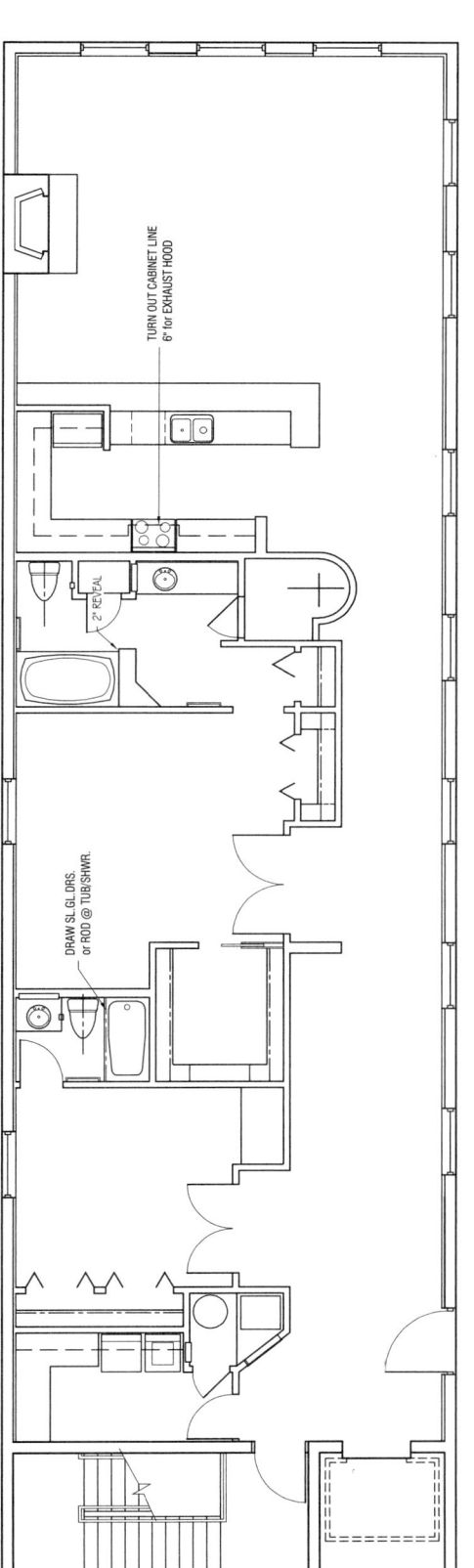

TURN OUT CABINET LINE
6" for EXHAUST HOOD

2" REVEAL

DRAW SL. GL. DRS.
or ROD @ TUB/SHWR.

INSERT BLOCKS -- (from CD Blocks 2000 or other):
TUBS, WC's, SINKS, LAVS. etc. -- PLUMBING LAYER
ELECTRICAL APPLIANCES        -- APPLIANCE  LAYER
ELEC. PANEL, H.W.T.          -- ELEC. EQUIP. LAYER
P.H.; M.C.; T.B.; HOOKS; ETC. -- EQUIPMENT LAYER

DRAW MIRRORS (OFFSET 1")  -- EQUIPMENT LAYER
INSERT FIREPLACE          -- FIREPLACE LAYER
DRAW AIR HANDLER (24x28), R.A.G. (30" WIDE) -- MECH. LAYER

BROWNSTONE: UPPER LEVEL APPLIANCES, FIXTURES, EQUIPMENT, FIREPLACE

---

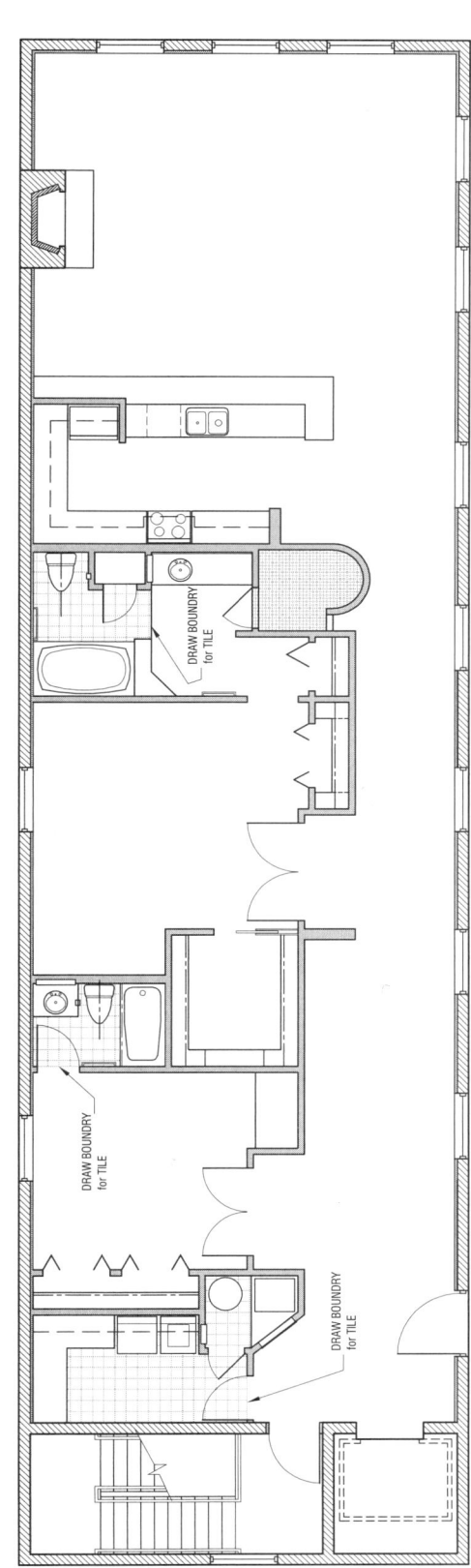

DRAW BOUNDRY
for TILE

DRAW BOUNDRY
for TILE

DRAW BOUNDRY
for TILE

HATCH MASONRY WALLS: ANSI31; SCALE FACTOR 24
HATCH FIRE BRICK: ANSI31; SCALE FACTOR 12, ANGLE 90
HATCH CER. TILE: "NET" PATTERN -- SCALE 64 TYP.; SCALE 12 @ SHOWER
                SEND TO BACK (TOOLS, DISPLAY ORDER)

HATCH SOLID (INTERIOR WALLS);  SEND to BACK (TOOLS, DISPLAY ORDER)

SAVE YOUR DRAWING BEFORE HATCHING!!
SELECT PATTERN from "ANSI" or "OTHER PREDEFINED"
WHEN in DOUBT, BEGIN with a SCALE FACTOR of 24 --
    ADJUST SCALE UP or DOWN from THERE.

BROWNSTONE: UPPER LEVEL BATHRM. EQUIP., HATCHING

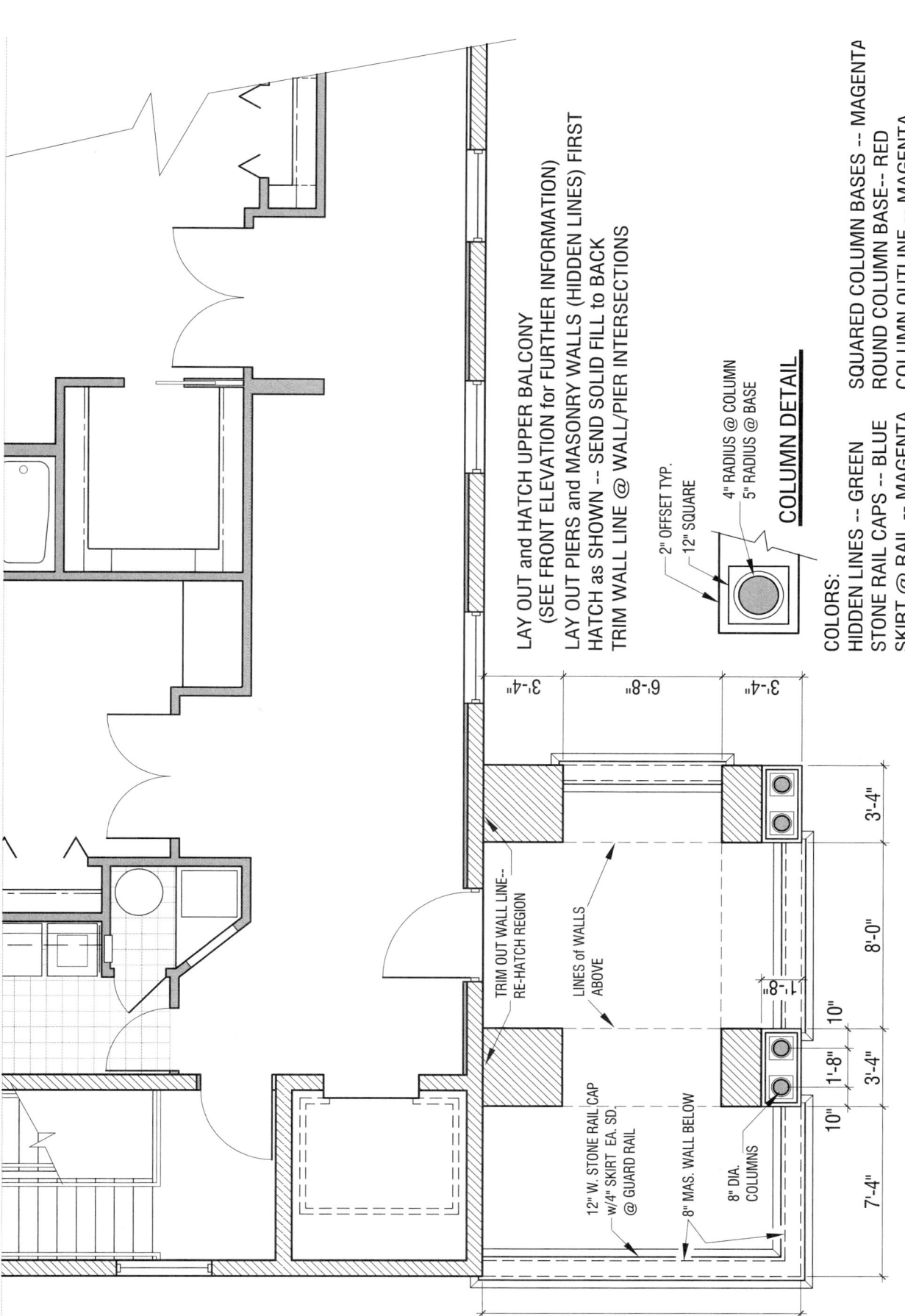

LAY OUT and HATCH UPPER BALCONY
(SEE FRONT ELEVATION for FURTHER INFORMATION)

LAY OUT PIERS and MASONRY WALLS (HIDDEN LINES) FIRST

HATCH as SHOWN -- SEND SOLID FILL to BACK
TRIM WALL LINE @ WALL/PIER INTERSECTIONS

2" OFFSET TYP.
12" SQUARE
4" RADIUS @ COLUMN
5" RADIUS @ BASE

COLUMN DETAIL

COLORS:
HIDDEN LINES -- GREEN          SQUARED COLUMN BASES -- MAGENTA
STONE RAIL CAPS -- BLUE        ROUND COLUMN BASE-- RED
SKIRT @ RAIL -- MAGENTA        COLUMN OUTLINE -- MAGENTA

TRIM OUT WALL LINE--
RE-HATCH REGION

LINES of WALLS
ABOVE

12" W. STONE RAIL CAP
w/4" SKIRT EA. SD.
@ GUARD RAIL

8" MAS. WALL BELOW

8" DIA.
COLUMNS

3'-4"
6'-8"
3'-4"

3'-4"
8'-0"
1'-8"
10"
1'-8"
3'-4"
10"
7'-4"

13'-4"

BROWNSTONE:  BALCONY/TOWER @ UPPER LEVEL

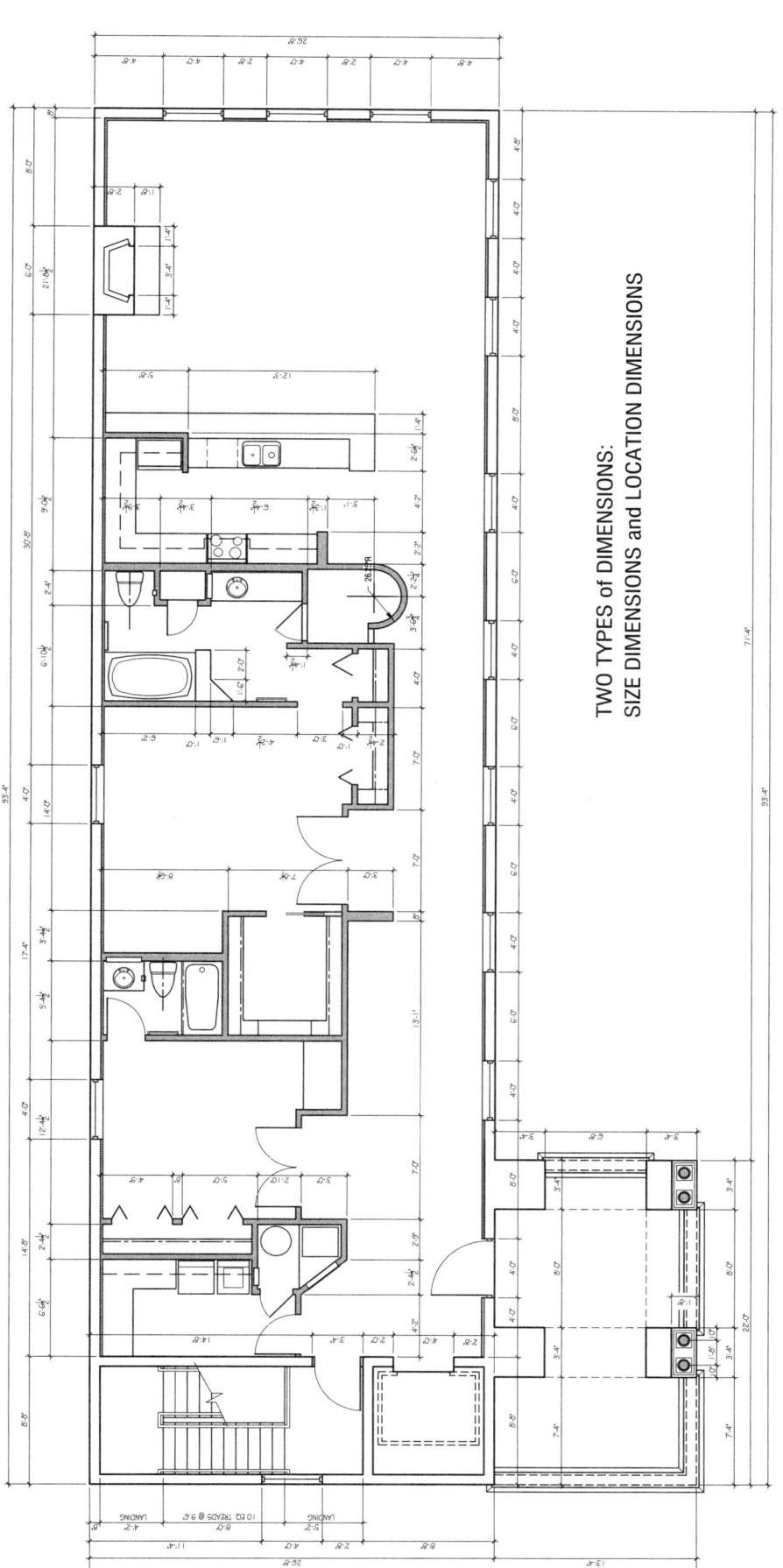

PLACE DIMENSIONS PER EXAMPLE -- SEE LARGE SCALE PLAN.
SET ARCHITECTURAL DIMENSIONING STYLE
USE OSNAPS and CONTINUE DIMENSION COMMAND
YOU MAY WISH to TURN OFF HATCH LAYERS to EASE SNAPPING
MAKE SURE to TURN ALL LAYERS BACK "ON" BEFORE MOVING PLAN!

TWO TYPES of DIMENSIONS:
SIZE DIMENSIONS and LOCATION DIMENSIONS

SYSTEM: LOCATE FIVE TYPES of STRINGS
1) OVERALL DIMENSION -- ALL SIDES
2) EXTERIOR JOGS NEXT (AS OCCUR)
3) EXTERIOR OPENINGS-- DOORS and WINDOWS
   WOOD FRAME -- DIM to CENTERLINE of OPENINGS
   MASONRY -- DIM to JAMBS (SIDES) of OPENINGS
4) INTERIOR PARTITIONS THAT ABUT EXTERIOR WALLS
5) REMAINING INTERIOR PARTITIONS and FEATURES

BROWNSTONE: UPPER LEVEL DIMENSIONS

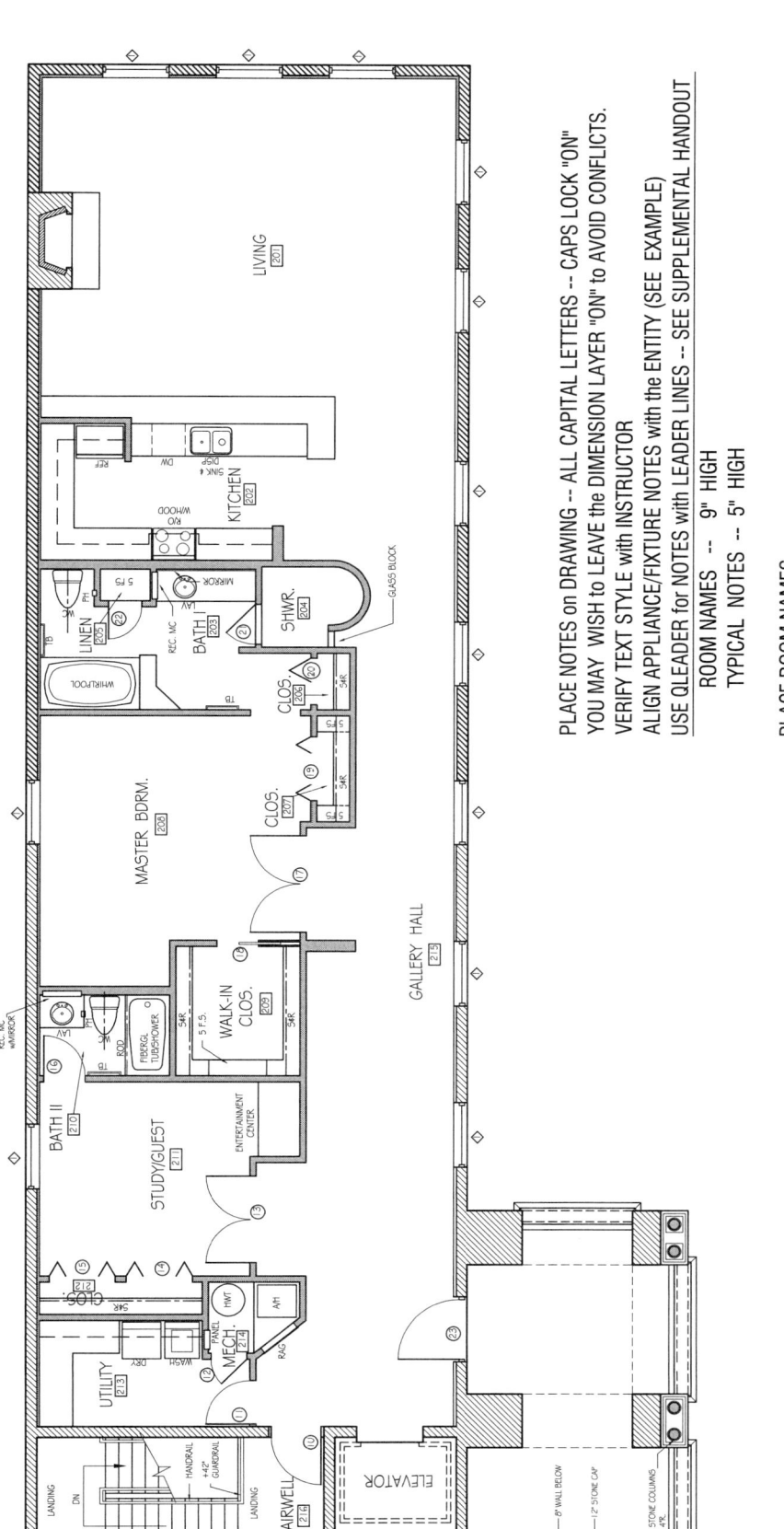

PLACE NOTES on DRAWING -- ALL CAPITAL LETTERS -- CAPS LOCK "ON"
YOU MAY WISH to LEAVE the DIMENSION LAYER "ON" to AVOID CONFLICTS.
VERIFY TEXT STYLE with INSTRUCTOR
ALIGN APPLIANCE/FIXTURE NOTES with the ENTITY (SEE EXAMPLE)
USE QLEADER for NOTES with LEADER LINES -- SEE SUPPLEMENTAL HANDOUT

    ROOM NAMES  --  9" HIGH
    TYPICAL NOTES -- 5" HIGH

PLACE ROOM NAMES
    DTEXT/CENTER JUSTIFY a "FAT NUMBER" EX: 888
    ENCLOSE with RECTANGLE then MULTIPLE COPY @ ALL ROOMS
    EDIT (ED) to CORRECT ROOM NUMBERS
PLACE DOOR and WINDOW NUMBERS -- SIMILAR PROCEDURE
PLACE TITLE and SCALE
    UNDERLINE W/ 1.5" POLYLINE (DRAW MENU)
    PICK START/ TYPE W (ENTER)
    ENTER START and END WIDTHS

BROWNSTONE: UPPER LEVEL
SCALE: 1/4" = 1'-0"

BROWNSTONE: ANNOTATION

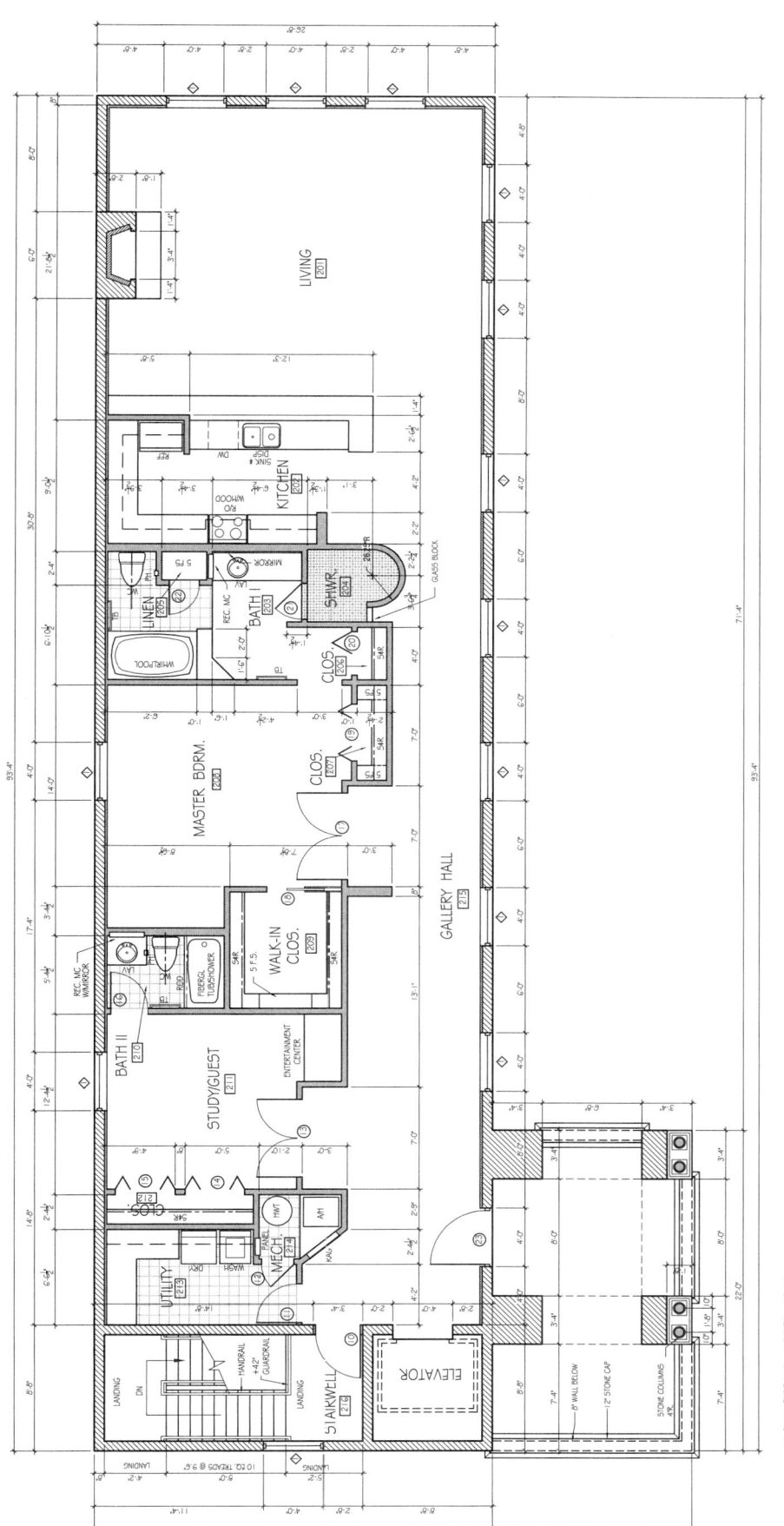

PLOT FINISHED UPPER LEVEL PLAN IN PAPER SPACE.
REFER TO QUICK GUIDE --VERIFY PERAMETERS W/INSTRUCTOR.
SELECT PLOTTING DEVICE for LARGE PLOT -- 24x36
SET PLOT STYLE for MONOCHROME, EDIT LINEWEIGHTS for LARGE PLOT
SET 24x36 PAPER for LANDSCAPE ORIENTATION

SCALE VIEWPORT for 1/4" - 1'-0"
PLACE VIEWPORT ON VIEWPORT LAYER (NO PLOT)

INSERT TITLE BLOCK -- REFERENCE INSERTION POINT TO (0,0)
COMPLETE TITLE BLOCK INFORMATION USING DTEXT IN PAPER SPACE
(USE "ACTUAL" HEIGHTS for TEXT -- $\frac{3}{32}$ or $\frac{1}{8}$; $\frac{3}{16}$ or $\frac{1}{4}$; etc.)

BROWNSTONE: UPPER LEVEL
SCALE: 1/4" = 1'-0"

BROWNSTONE:  FINISHED UPPER LEVEL PLAN

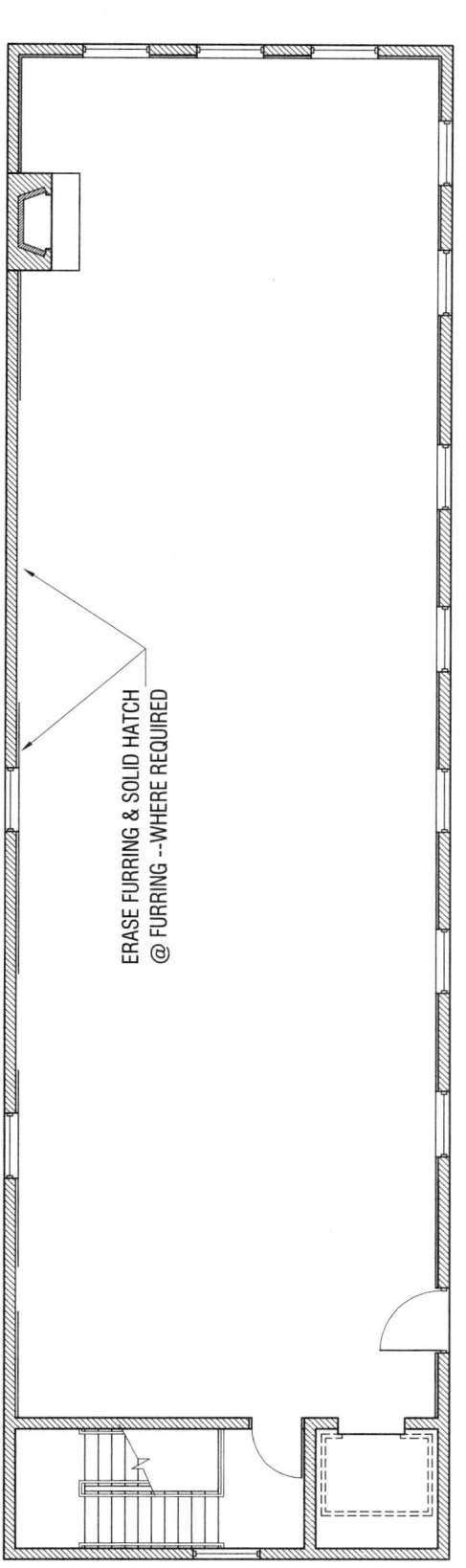

ERASE FURRING & SOLID HATCH
@ FURRING --WHERE REQUIRED

TO BEGIN the LOWER LEVEL .....
COPY the COMPLETED UPPER LEVEL PLAN to the LOWER PART of YOUR SCREEN -- ORTHO "ON".
ERASE ENTITIES THAT APPLY to the UPPER LEVEL ONLY
    USE CROSSING WINDOWS for SPEEDY SELECTION.....
THE RESULT SHOULD LOOK LIKE the EXAMPLE ABOVE.

BROWNSTONE: LOWER LEVEL BLOCK OUT

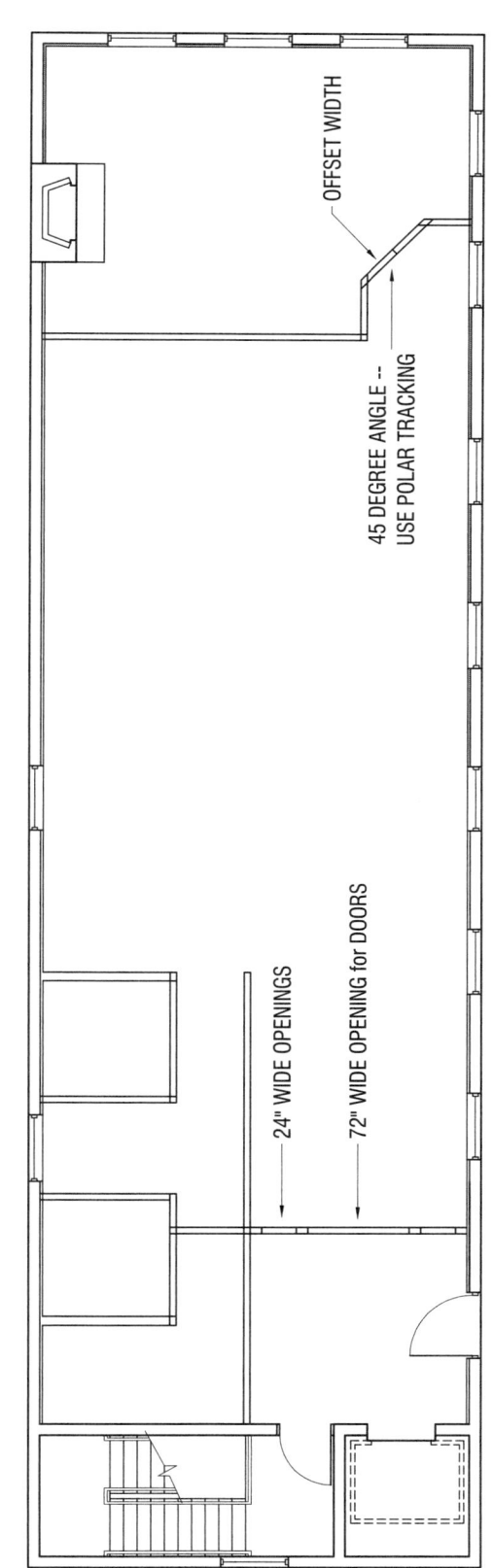

OFFSET WIDTH

45 DEGREE ANGLE --
USE POLAR TRACKING

24" WIDE OPENINGS

72" WIDE OPENING for DOORS

OFFSET EXTERIOR WALLS to LOCATE INTERIOR PARTITIONS
DO a "ROUGH TRIM" to DEFINE WALLS
MOVE to WALLS_INT LAYER

YOU MAY WISH to TURN "OFF" the MASONRY HATCH LAYER
DO NOT "MOVE" the PLAN WHILE LAYERS are TURNED "OFF" --
    THE "OFF" LAYERS WILL BE LEFT BEHIND.....MAKES A REAL MESS!

BROWNSTONE: LOWER LEVEL INTERIOR PARTITIONS

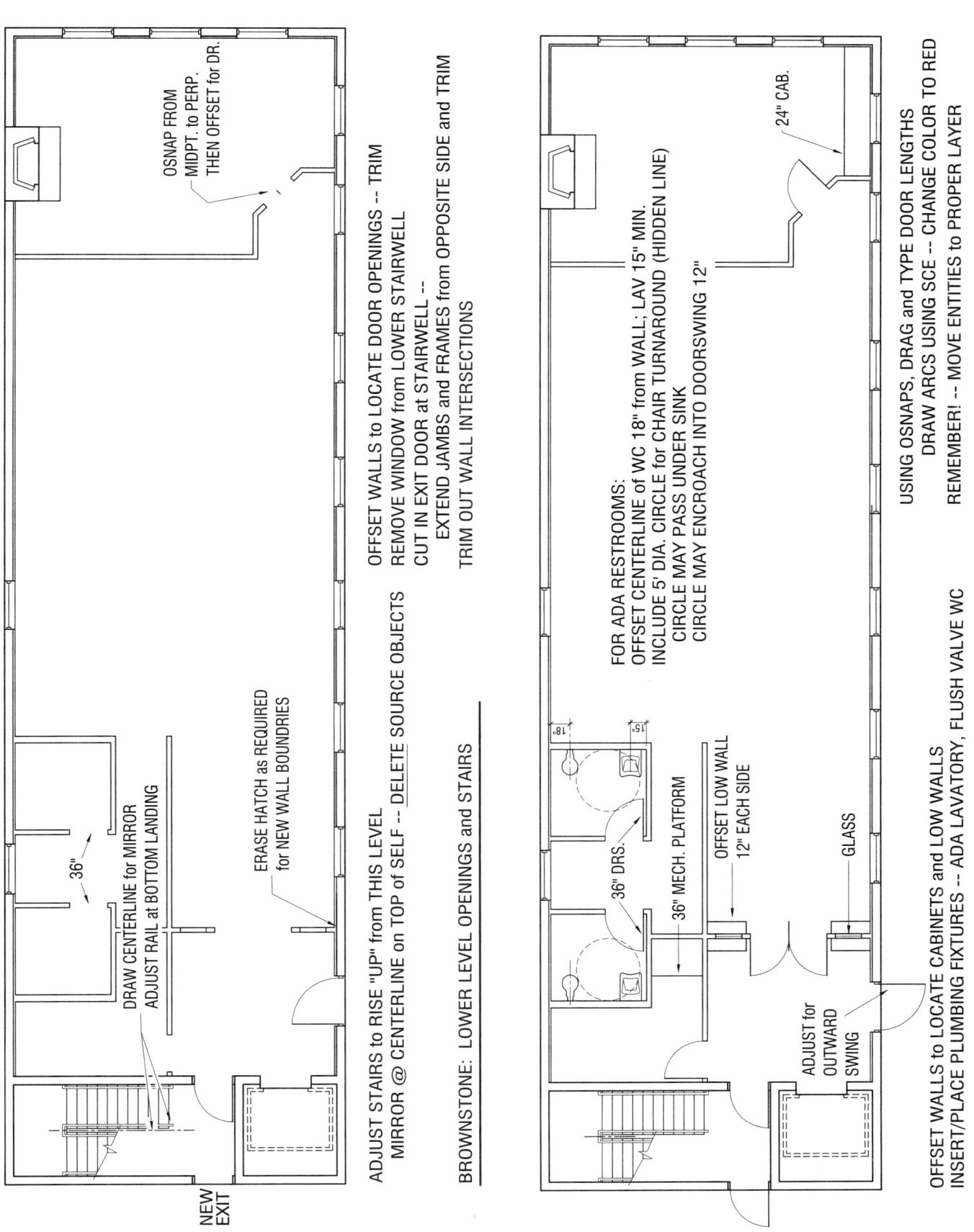

OSNAP FROM
MIDPT. to PERP.
THEN OFFSET for DR.

OFFSET WALLS to LOCATE DOOR OPENINGS -- TRIM
REMOVE WINDOW from LOWER STAIRWELL --
CUT IN EXIT DOOR at STAIRWELL --
EXTEND JAMBS and FRAMES from OPPOSITE SIDE and TRIM
TRIM OUT WALL INTERSECTIONS

36"

DRAW CENTERLINE for MIRROR
ADJUST RAIL at BOTTOM LANDING

ERASE HATCH as REQUIRED
for NEW WALL BOUNDRIES

NEW
EXIT

ADJUST STAIRS to RISE "UP" from THIS LEVEL
MIRROR @ CENTERLINE on TOP of SELF -- DELETE SOURCE OBJECTS

BROWNSTONE:  LOWER LEVEL OPENINGS and STAIRS

24" CAB.

USING OSNAPS, DRAG and TYPE DOOR LENGTHS
DRAW ARCS USING SCE -- CHANGE COLOR TO RED
REMEMBER! -- MOVE ENTITIES to PROPER LAYER

FOR ADA RESTROOMS:
OFFSET CENTERLINE of WC 18" from WALL; LAV 15" MIN.
INCLUDE 5' DIA. CIRCLE for CHAIR TURNAROUND (HIDDEN LINE)
CIRCLE MAY PASS UNDER SINK
CIRCLE MAY ENCROACH INTO DOORSWING 12"

18"

15"

36" DRS.

36" MECH. PLATFORM

OFFSET LOW WALL
12" EACH SIDE

GLASS

ADJUST for
OUTWARD
SWING

OFFSET WALLS to LOCATE CABINETS and LOW WALLS
INSERT/PLACE PLUMBING FIXTURES -- ADA LAVATORY, FLUSH VALVE WC

BROWNSTONE:  LOWER LEVEL CABINETS, DOORS, PLUMBING FIXTURES

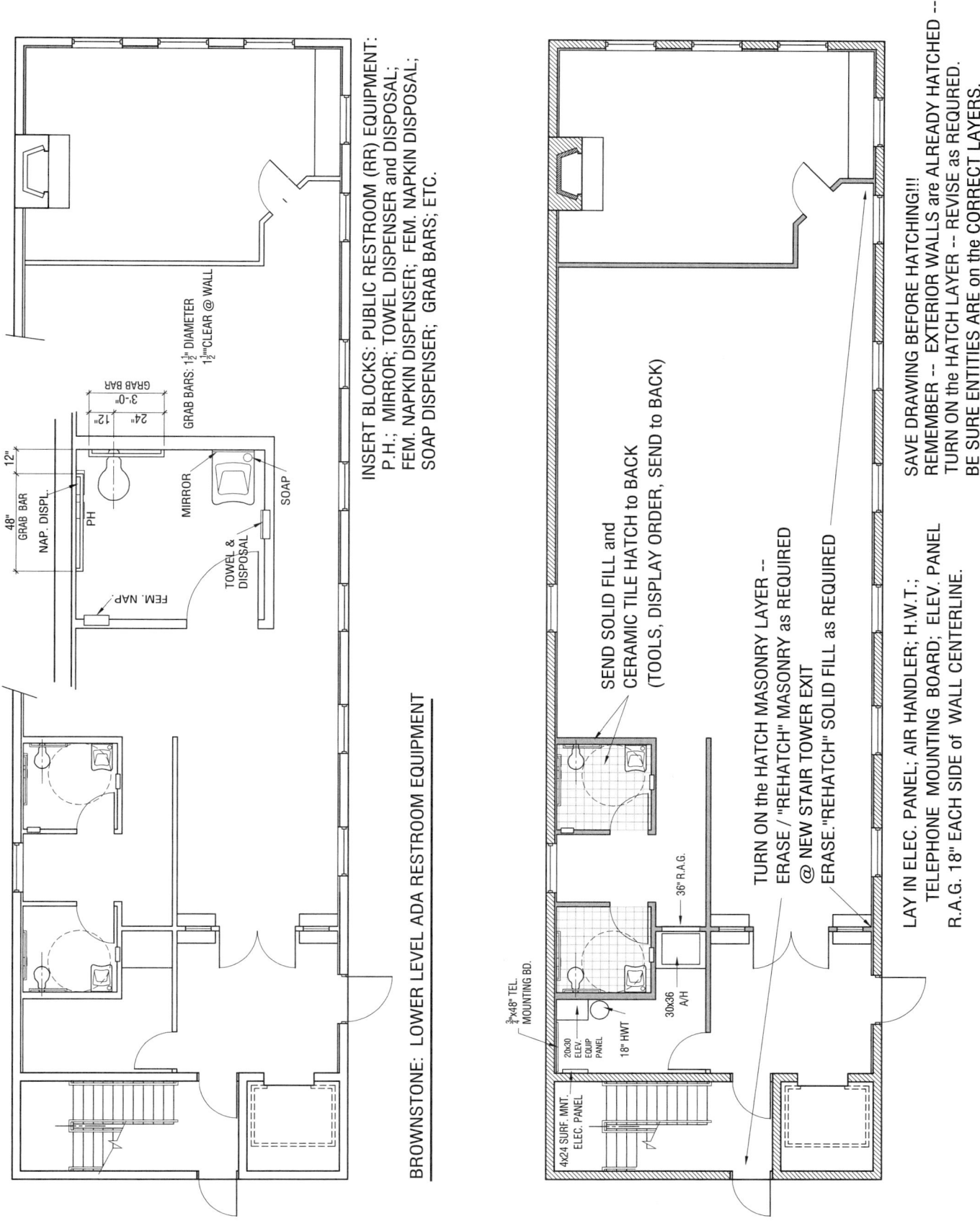

INSERT BLOCKS: PUBLIC RESTROOM (RR) EQUIPMENT:
P.H.; MIRROR; TOWEL DISPENSER and DISPOSAL;
FEM. NAPKIN DISPENSER; FEM. NAPKIN DISPOSAL;
SOAP DISPENSER; GRAB BARS; ETC.

GRAB BARS: 1½" DIAMETER
1½" CLEAR @ WALL

GRAB BAR
3'-0"
24"  12"

12"

48"
GRAB BAR
NAP. DISPL.

PH

MIRROR

SOAP

FEM. NAP.

TOWEL &
DISPOSAL

BROWNSTONE:  LOWER LEVEL ADA RESTROOM EQUIPMENT

SEND SOLID FILL and
CERAMIC TILE HATCH to BACK
(TOOLS, DISPLAY ORDER, SEND to BACK)

TURN ON the HATCH MASONRY LAYER --
ERASE / "REHATCH" MASONRY as REQUIRED
@ NEW STAIR TOWER EXIT
ERASE "REHATCH" SOLID FILL as REQUIRED

SAVE DRAWING BEFORE HATCHING!!!
REMEMBER --  EXTERIOR WALLS are ALREADY HATCHED --
TURN ON the HATCH LAYER -- REVISE as REQUIRED.
BE SURE ENTITIES ARE on the CORRECT LAYERS.

36" R.A.G.

30x36
A/H

18" HWT

20x30
ELEV.
EQUIP
PANEL

¾"x48" TEL.
MOUNTING BD.

4x24 SURF. MNT.
ELEC. PANEL

LAY IN ELEC. PANEL; AIR HANDLER; H.W.T.;
TELEPHONE MOUNTING  BOARD;  ELEV. PANEL
R.A.G. 18" EACH SIDE of  WALL CENTERLINE.

BROWNSTONE:  LOWER LEVEL MECH/ELEC EQUIPMENT;  HATCHING

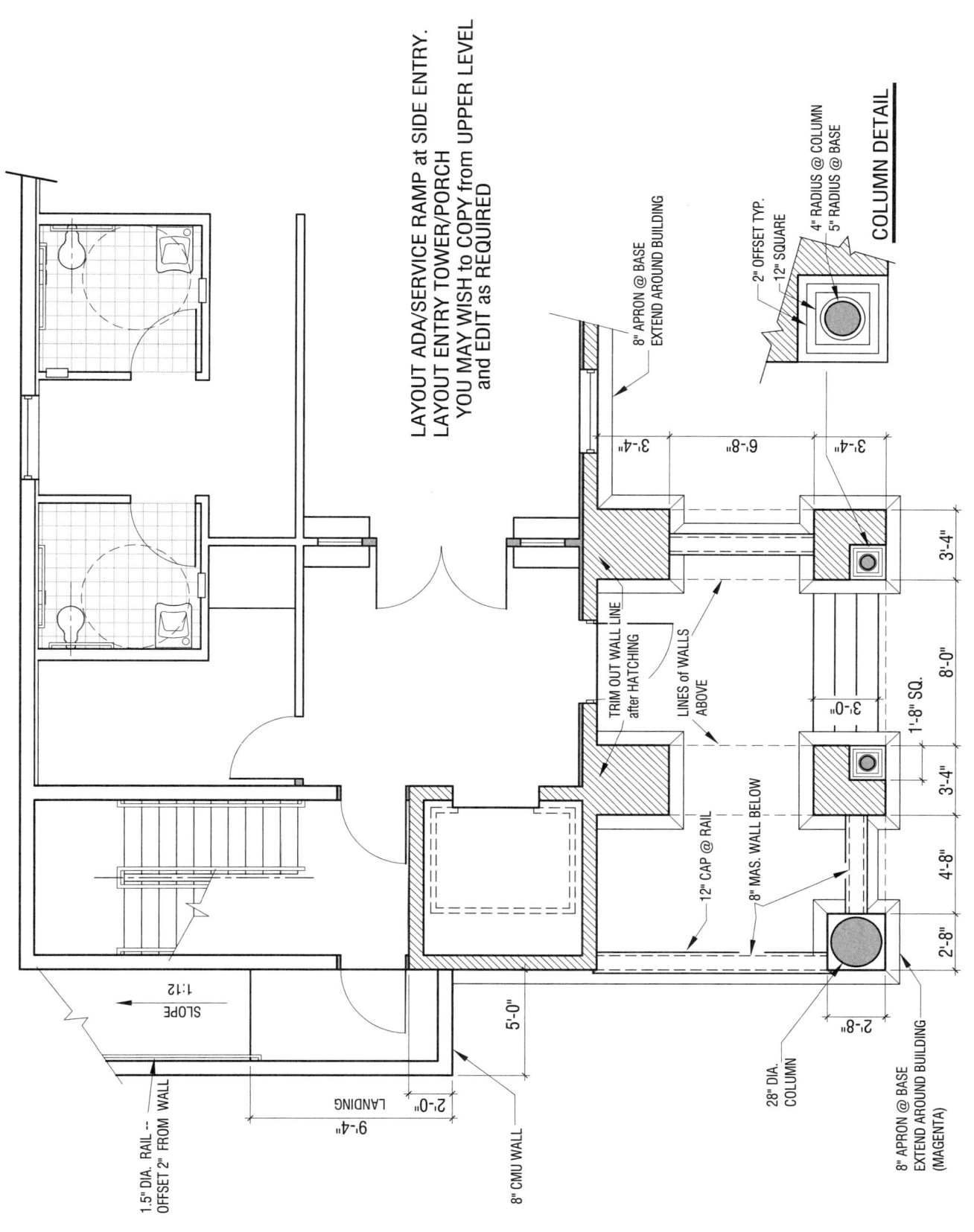

LAYOUT ADA/SERVICE RAMP at SIDE ENTRY.
LAYOUT ENTRY TOWER/PORCH
YOU MAY WISH to COPY from UPPER LEVEL
and EDIT as REQUIRED

COLUMN DETAIL

4" RADIUS @ COLUMN
5" RADIUS @ BASE
2" OFFSET TYP.
12" SQUARE

8" APRON @ BASE
EXTEND AROUND BUILDING

3'-4"
6'-8"
3'-4"

3-4"
8'-0"
3-4"
2'-8"

1'-8" SQ.
3'-0"
4'-8"

TRIM OUT WALL LINE
after HATCHING

LINES of WALLS
ABOVE

12" CAP @ RAIL

8" MAS. WALL BELOW

28" DIA.
COLUMN

8" APRON @ BASE
EXTEND AROUND BUILDING
(MAGENTA)

2'-8"

8" CMU WALL

2'-0" LANDING

9'-4"

5'-0"

SLOPE 1:12

1.5" DIA. RAIL --
OFFSET 2" FROM WALL

BROWNSTONE:  LOWER  LEVEL -- PORCH and RAMP

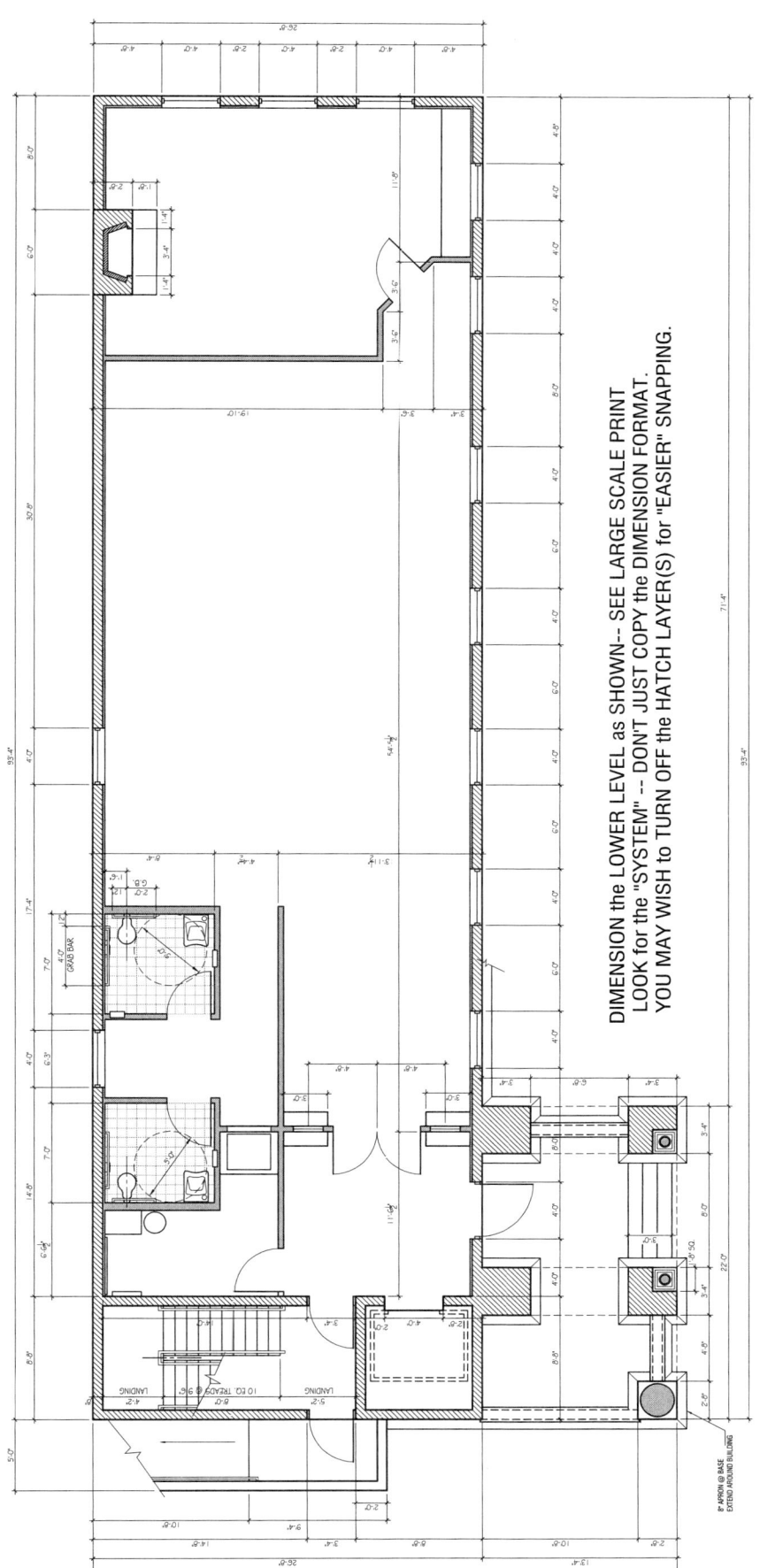

DIMENSION the LOWER LEVEL as SHOWN-- SEE LARGE SCALE PRINT
LOOK for the "SYSTEM" -- DON'T JUST COPY the DIMENSION FORMAT.
YOU MAY WISH to TURN OFF the HATCH LAYER(S) for "EASIER" SNAPPING.

BROWNSTONE: LOWER LEVEL -- DIMENSIONS

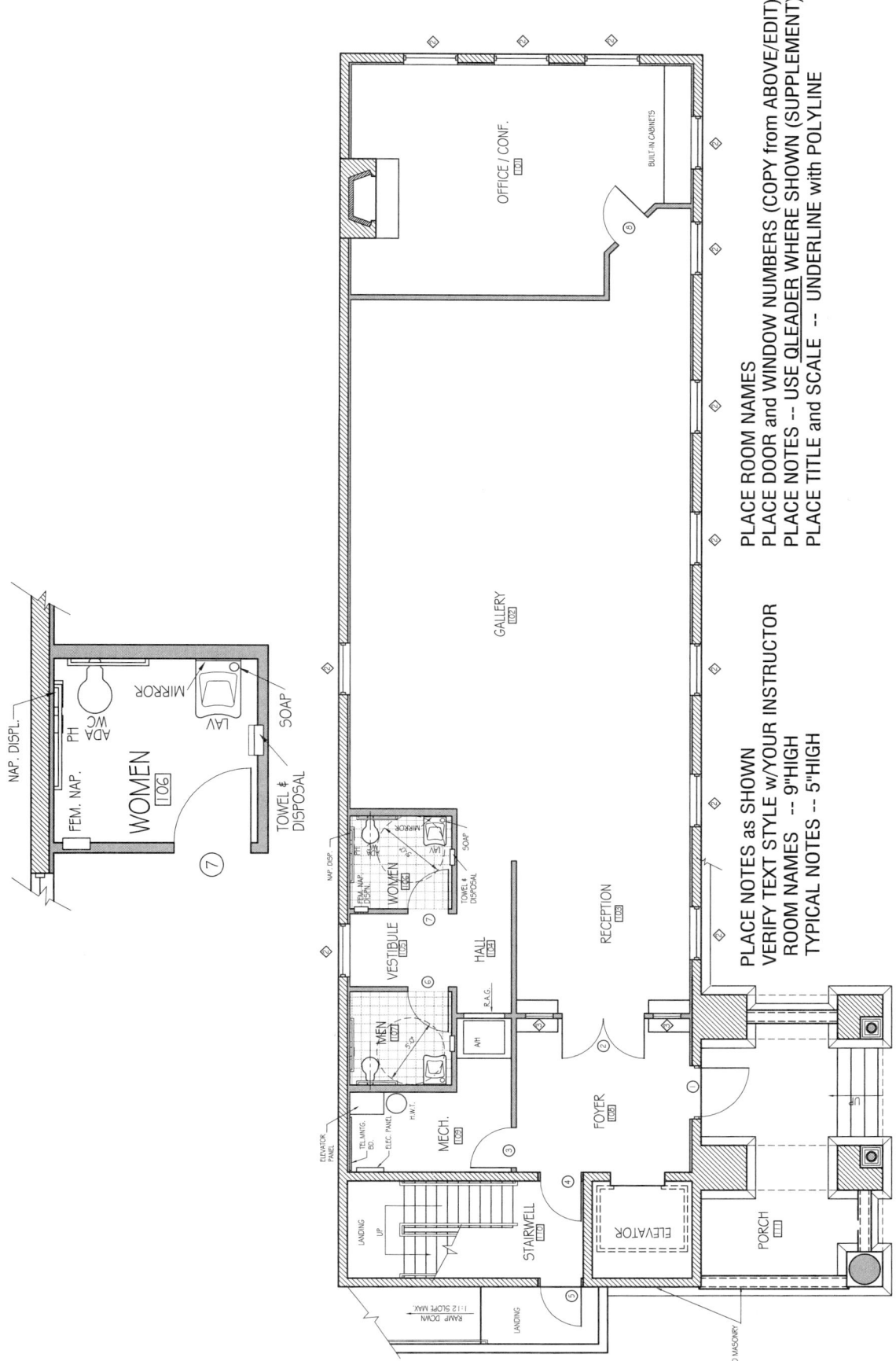

PLACE ROOM NAMES
PLACE DOOR and WINDOW NUMBERS (COPY from ABOVE/EDIT)
PLACE NOTES -- USE QLEADER WHERE SHOWN (SUPPLEMENT)
PLACE TITLE and SCALE -- UNDERLINE with POLYLINE

PLACE NOTES as SHOWN
VERIFY TEXT STYLE w/YOUR INSTRUCTOR
ROOM NAMES -- 9"HIGH
TYPICAL NOTES -- 5"HIGH

OFFICE / CONF.

BUILT-IN CABINETS

GALLERY

RECEPTION

WOMEN

NAP. DISPL.

FEM. NAP.
PH
ADA
WC
MIRROR
LAV
SOAP
TOWEL &
DISPOSAL

NAP DISP
FEM. NAP. DISPL.
WOMEN
MIRROR
LAV
SOAP
TOWEL &
DISPOSAL

VESTIBULE
HALL
R.A.G.
AH
MEN

ELEVATOR PANEL
TEL.MNTG.
BD
ELEC. PANEL
H.W.T.
MECH.

FOYER

LANDING
UP
STAIRWELL

RAMP DOWN
1:12 SLOPE MAX.
LANDING

EXPOSED MASONRY
TYPICAL

ELEVATOR

UP

PORCH

BROWNSTONE: LOWER LEVEL
SCALE: 1/4" = 1'-0"

BROWNSTONE:  LOWER  LEVEL  --  ANNOTATION

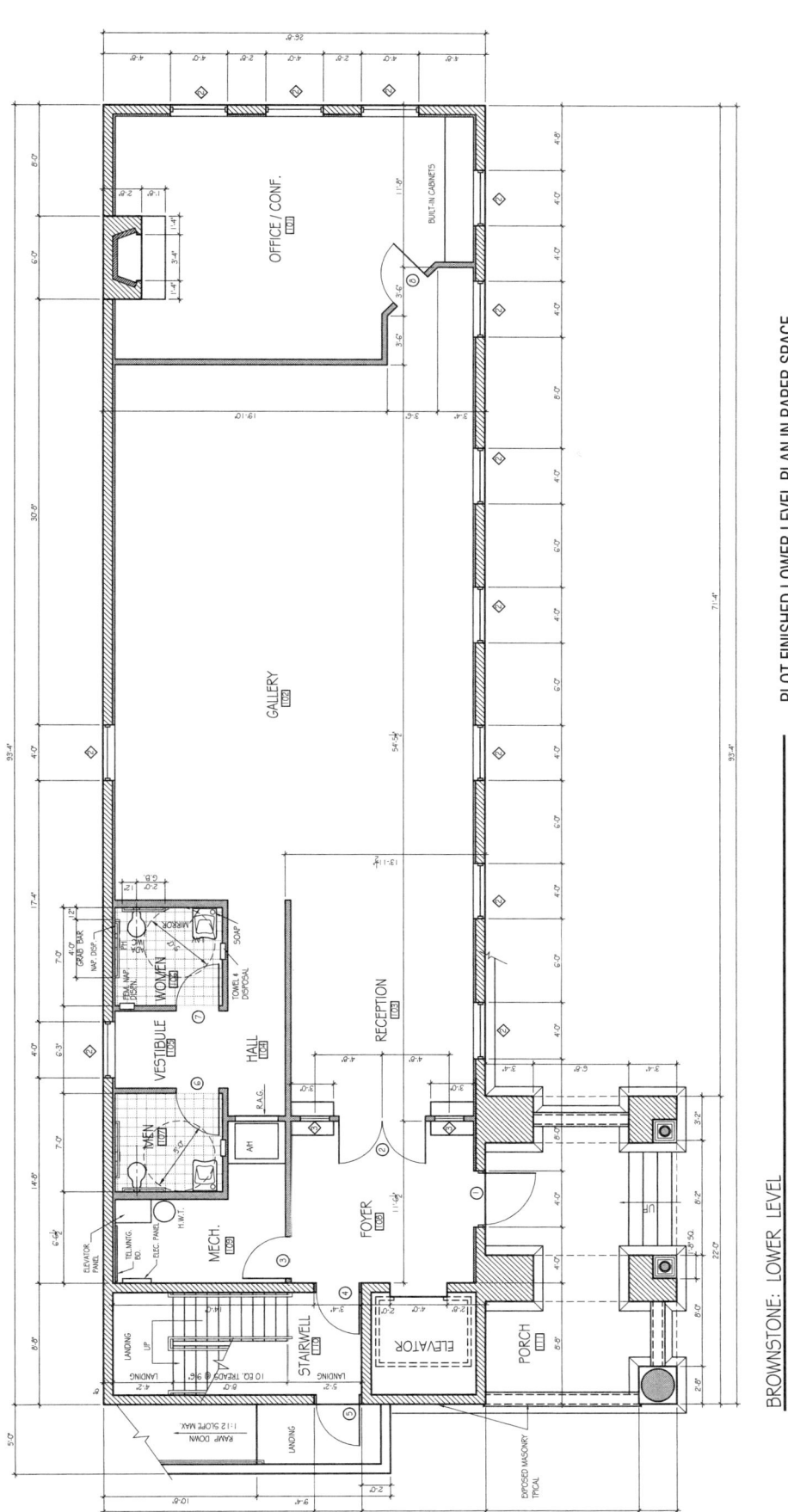

PLOT FINISHED LOWER LEVEL PLAN IN PAPER SPACE.
    REFER TO QUICK GUIDE --VERIFY PERAMETERS W/INSTRUCTOR.
SELECT PLOTTING DEVICE for LARGE PLOT -- 24x36
SET PLOT STYLE for MONOCHROME, EDIT LINEWEIGHTS for LARGE PLOT
SET 24x36 PAPER for LANDSCAPE ORIENTATION

SCALE VIEWPORT for 1/4" - 1'-0"
PLACE VIEWPORT ON VIEWPORT LAYER (NO PLOT)

INSERT TITLE BLOCK -- REFERENCE INSERTION POINT TO (0,0)
COMPLETE TITLE BLOCK INFORMATION USING DTEXT IN PAPER SPACE
    (USE "ACTUAL" HEIGHTS for TEXT -- $\frac{3}{32}$ or $\frac{1}{8}$; $\frac{3}{16}$ or $\frac{1}{4}$; etc.)

BROWNSTONE: LOWER LEVEL

BROWNSTONE: FINISHED LOWER LEVEL PLAN

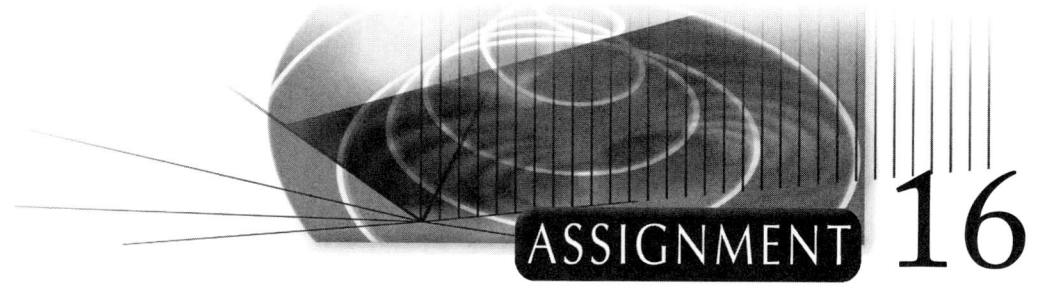

# Exterior Elevation(s)

Follow the illustrations and notes in laying out the Front Exterior Elevation of the Furness Brownstone. Verify the "order" in which you layout components with your instructor. While basic dimensions are given, you will need to do some interpolating to complete some components. Size the nondimensioned items by analyzing/comparing to other, dimensioned items. You should make good use of the Copy and Mirror commands—to save your time, **make sure that you adjust Colors for Plotting BEFORE you Copy, Multiple Copy, and/or Mirror.** You may wish to run a check plot to verify Lineweights before you start replicating images.

Recall the methodology employed in the Georgian House for placing windows.

### SAVE ELEVATION WORK TO YOUR DESIGNATED DRIVE; BACKUP TO A DISK

**Verify with your instructor:** Is this assignment (16) a "New drawing"?

Or a continuation of Assignments 14 and 15?

**If continuing in the same drawing file** as the Upper and Lower Level Plans:

**Create New Layers:** Elevation (White)
Text_elev (Yellow)
Dim_elev (Red) Text Color is set in Dimension Style.

And off you go…

You may wish to project widths from your plan views to the Elevation, turning off some layers might make things easier

**If setting up a New Drawing file:**

**Use a Wizard**

**Units: Architectural**

Pick **Next**

**Area: 144' × 96'** (limits for ¼" scale on a 36" × 24" paper—see *Supplement 2: Drawing Area*)

Pick **Finish**

**Zoom All** to capture your full screen.

**Save As** to name this drawing—your three initials BRNELEV Example: MRB BRNELEV

**Load Linetypes:** Center; **Global Linetype Scale** = 24

**Create New Layers:** Elevation (White)
Text_elev (Yellow)
Dim_elev (Red) Text Color is set in Dimension Style

## PROCEDURE

Refer to drawings and instructions for the Windows, Tower, and overall Elevation. You will make many decisions on your own over the course of this drawing. Use your eye and best judgment in spacing elements and interpolating nondimensioned entities by comparing them to dimensioned items.

**Verify your starting point with your instructor**—your assignments may hinge on semester scheduling. Some students begin by drawing the Windows, others begin with the Tower, others begin with the overall block out (as you typically would in an office situation). Have some fun with this. Pay special attention to Lineweights (Colors) and you'll end up with a great looking $1/4$" scale 36 × 24 Plot, or using small Lineweights settings, an exquisite reduced scale Plot.

**MORE WORK?!** At the direction of your instructor, complete the Right Side, Rear, and Left Side Elevations, "figuring out" what they look like from Plan and Front views.

**Plot in Paper Space when you're finished, with a Title Block**

      **WOW!... take a minute to reflect on where you began and the work you've completed.**

SUPPLEMENTS

SUPPLEMENT 1

# General Operations

## GENERAL AUTOCAD OPERATIONS

The following information gives a review of General Operations for AutoCAD, including **Accessing Commands and Toolbars, Picking or Selecting Entities, Undo and Escape, and Drawing and Measuring Lines.**

**Accessing Commands in AutoCAD:**

Most AutoCAD commands may be accessed in a variety of ways including Pulldown Menus, Toolbar Icons, Typing and Entering the command name, or Typing and Entering a command Alias. The Alias is typically the first or first two letters of the command, i.e., the Alias for the Line command is **L**, the Alias for the Erase command is **E.**

When searching for a command, think in general terms. If you want to create something new, "out of thin air," then you most probably want a **Draw** command. That would direct your search to the **Draw Pulldown** or **Draw Toolbar.** Likewise, if you want to change or build on what is already in place, "modify what's in place," then searching the **Modify Pulldown** or **Modify Toolbar** would be a good first step. If your search proves fruitless, you can always try "typing" the command (and Entering) or an Alias.

And finally, don't be shy about using AutoCAD's Help Directory—the icon with the question mark (or just "type" (and Enter) **Help**).

**Picking and Entering:**

The key to learning AutoCAD is to **Read the Command Prompt Line.** It will walk you through almost every command (and will actually begin to make real sense as your experience grows). Many of the operations involve Picking or Selecting Entities and Entering when you're finished.

To **Pick or Select** commands and to **Select Objects** on your drawing screen "one at a time":

> **Left Click** with the mouse.

To **Pick or Select** *groups* of Objects—use a Pick Window or Crossing Window:

> A Window built left to right (Pick Window) will select *only* those items *completely* inside it.

> A Window built right to left (Crossing Window) will select items inside it **and** entities it *crosses.*

>> You typically build these Windows by **left clicking** two points or corners. Pick the first point on empty screen—if you're "on" an Object, only that Object will be selected.

**Enter** to tell AutoCAD that you're **"finished"**—with a command, typing, or selecting Objects:

> Use the **Enter** key on keyboard—at keypad or number pad

> **Right Click** on Mouse

> **Space Bar** on Keyboard

>> Entering (try right clicking) a second time, immediately after entering to finish a command, will repeat the command—unless AutoCAD is defaulting to the Shortcut Menu.

The Erase command offers good practice here, you can select Objects individually, by picking them one by one, select additional Objects in a Pick Window, and select still more in a Crossing Window—picking away until you're through. Hit **Enter** to tell AutoCAD you're done selecting. The Pick Window is a great way to capture a small entity within a larger group—it will only grab things *completely* inside it, right?

### Undo, Redo, and Escape:

The **Undo** command is a useful tool. You can use it to bring back entities you've inadvertently lost, or to get rid of work that didn't turn out quite like you had hoped. You can "**Undo**" through your days' commands in reverse order. The related command, **Redo,** will reclaim your "**Undo's**" in reverse order. Caution here, though. In versions older than 2004, the **Redo** command will let you "reclaim" only ONE Undo.

**Escape** (upper left on keyboard) will cancel the current command—great for getting yourself out of the middle of some command mess and starting over with a clean Command Prompt Line.

### Toolbars:

A wide variety of Toolbars are available to help you on your way, but which Toolbars are displayed on the screen is often a matter a personal preference.

To **Load** a Toolbar, Right Click on any Toolbar and the complete Toolbar list will appear. Select the one(s) you're after. Typically they will appear in the drawing area. You can move a Toolbar by left clicking and dragging in the colored area at the top (probably blue). If you drag it above, below, or to the side of your screen, it will reconfigure to vertical or horizontal and **Dock** when you release the left mouse button.

To **Close** a Toolbar, Left Click on the X at the colored upper area. To **Close** or **Move** a **Docked** Toolbar, pick and drag (left click) between the two (light) gray lines at the end of the bar. This will allow you to shuffle the positions, or drag it to the screen and "**X**" it away.

More Toolbar options are available through the **View** Pulldown/**Toolbars/Customize.** Through this route you can select individual icons for your screen or modify typical Toolbar configurations.

### Drawing and Measuring Lines:

Well, this is what we're talking about—drawing lines—right?

Accessing the Line command—four ways:

- Select from **Draw** Pulldown/Line (left click).
- Pick the **Line** icon from the **Draw Toolbar** (left click).
- Type **Line** and then **Enter**
- Type **L** and then **Enter** (**L** is the Alias for **L**ine). Remember that the Space Bar works as an Enter option. This speeds up "typed commands."

Follow the prompts, picking a starting point and then subsequent points by left clicking the mouse.

AutoCAD will continue to ask you for "next points" untill you're dead and buried. Only you know when you want to stop drawing. To tell AutoCAD you're done—**Enter** (Right Click or Enter key). This process will carry through for many commands, i.e., you **MUST** tell AutoCAD when you're done typing, done selecting objects, etc., by **Entering.**

A quick and efficient way to draw rectilinear objects is to "**Drag, Type, and Enter.**" In the **Line** command, Pick the "first point" with a Left Click, then, *with Ortho ON,* just **drag** the mouse in the direction you want to go. **Type** the distance (no dash or inch mark required) and **Enter.** The Line segment produced will fall in the direction you indicated and be the exact distance that you typed. You can draw continuous segments by repeating the drag/type/Enter process, working your way around an object. Just pick a starting point, then **drag/type/Enter** … **drag/type/Enter** … **drag/type/Enter** … and **Enter** to finish. Sometimes you must "wiggle" the mouse a bit to "find" your last corner. Make sure that you are "seeing" the correct direction of the line when you drag, before you **Enter** the distance.

To measure the accuracy of a line, use the **Distance** command (**Tools/Inquiry/Distance, Inquiry Toolbar,** or type **DI** and **Enter.** For accuracy, you *must* use OSnaps to measure entities with the **Distance** command. A related command is the **List** command, which tells you all you ever wanted to know about an entity, and probably a lot more than you ever wanted to know.

# Drawing Area/Limits

## DRAWING AREA/DRAWING LIMITS AND SCALE FACTORS

Sizing Your Sheet of Paper in Model Space

The following table shows Scale Factors, Drawing Area, and Drawing Limits for typical Architectural Scales and drawing/plotting paper sizes. Though all these items may be modified during the drawing process, **you will need to make a preliminary decision about both the *Scale* of your final Plot and the *Size* of paper on which you intend to Plot during your drawing Setup prior to using this table.**

NOTE! Values for **Drawing Area** and **Drawing Limits** assume a **Landscape** orientation—the "wide" side of the paper lying in the horizontal direction. To set up for **Portrait** orientation, **you must reverse** the width/length or the upper right limit coordinates.

Scale Factor Settings:

> Scale Factor = 12 (i.e., 1'-0") divided by the fractional scaled equivalent.
>
> Ex: $1/2$" = 1'-0" … 12 divided by $1/2$ = 24 (invert fraction and multiply).
>
> **Applications of Scale Factors:**
>
> **Linetype settings:** In **Format** Pulldown/**Linetype**
>
> > Set **Global Scale Factor** to "half" the Drawing Scale Factor.
> >
> > Ex: For $1/4$" = 1'-0" (Scale Factor 48) set Global Scale Factor = 24.
> >
> > (Some use the "actual" Scale Factor here.)
>
> > **Dimension Style:** In **Dimension Style/ FIT** tab
> >
> > > … **Use overall scale of** … use full scale factor.
> > >
> > > Ex: For $1/4$" = 1'-0" (Scale Factor 48) set FIT to 48.

Drawing Area: (Paper Size in Model Space)

> In **Wizard/Quick Setup** set widths and lengths of the Drawing Area
>
> in accordance with the Paper Size and the Scale of the final Plot.

Drawing Limits:

Drawing limits gauge Drawing Area using **(x,y)** coordinates, "**x**" representing a horizontal distance, "**y**" representing a vertical distance.

Think of your drawing area displayed on a Grid for 36 × 24 Plotted size at $1/4$" = 1'-0" scale. Referring to the chart, the corresponding Drawing Limits = (0,0) (144',96'). Coordinates (0,0) would be the bottom left corner of the Grid and (144',96') would locate the upper right corner—from the bottom left (0,0) you must travel 144' *across* and 96' *up* to move to the upper right corner of the Grid.

To verify your Drawing Limits go to **Format** pulldown/**Drawing Limits.** To modify the limits, typically accept (0,0) as the bottom left corner, and enter the desired upper right corner as a pair of (x,y) coordinates. Remember the comma between the two values.

**Scale Factors/Drawing Area/Drawing Limits per Paper Size and Plotted Scale**

| PAPER SIZE | PLOTTED SCALE | SCALE FACTOR | DRAWING AREA | DRAWING LIMITS |
|---|---|---|---|---|
| 11 × 8 ½ (letter) | 1/16 in. = 1' | 192 | 176' × 136' | 0,0 / 176', 136' |
| | 1/8 in. = 1' | 96 | 88' × 68' | 0,0 / 88', 68' |
| | 3/16 in. = 1' | 64 | 58'-8" × 45'-4" | 0,0 / 58'-8", 45'-4" |
| | 1/4 in. = 1' | 48 | 44' × 34'-4" | 0,0 /44', 34'-4" |
| | 3/8 in. = 1' | 32 | 29'-4" × 22'-8" | 0,0 / 29'-4", 22'-8" |
| | 1/2 in. = 1' | 24 | 22' × 17' | 0,0 / 22', 17' |
| | 3/4 in. = 1' | 16 | 14'-8" × 11'-4" | 0,0 /14'-8",11'-4" |
| | 1 in. = 1' | 12 | 11' × 8'-6" | 0,0 / 11', 8'-6" |
| | 1.5 in. = 1' | 8 | 7'-4" × 5'-8" | 0,0 / 7'-4", 5'-8" |
| | 1 in. = 20' | 240 | 220' × 170' | 0,0 /220', 170' |
| 17 × 11 (tabloid) | 1/16 in. = 1' | 192 | 272' × 176' | 0,0 / 272',176' |
| | 1/8 in. = 1' | 96 | 136' × 88' | 0,0 / 136', 88' |
| | 3/16 in. = 1' | 64 | 90'-8" × 58'-8" | 0,0 / 90'-8", 58'-8" |
| | 1/4 in. = 1' | 48 | 68' × 44' | 0,0 / 68', 44' |
| | 3/8 in. = 1' | 32 | 45'-4" × 29'-4" | 0,0 / 45'-4", 29'-4" |
| | 1/2 in. = 1' | 24 | 34' × 22' | 0,0 / 34', 22' |
| | 3/4 in. = 1' | 16 | 22'-8" × 14'-8" | 0,0 / 22'-8", 14'-8" |
| | 1 in. = 1' | 12 | 17' × 11' | 0,0 / 17', 11' |
| | 1.5 in. = 1' | 8 | 11'-4" × 7'-4" | 0,0 / 11'-4", 7'-4" |
| | 1 in. = 20' | 240 | 340' × 220' | 0,0 / 340', 220' |
| 36 × 24 | 1/16 in. = 1' | 192 | 576' × 384' | 0,0 / 576', 384' |
| | 1/8 in. = 1' | 96 | 288' × 192' | 0,0 / 288', 192' |
| | 3/16 in. = 1' | 64 | 192' × 128' | 0,0 / 192', 128' |
| | 1/4 in. = 1' | 48 | 144' × 96' | 0,0 / 144', 96' |
| | 3/8 in. = 1' | 32 | 96' × 64' | 0,0 / 96', 64' |
| | 1/2 in. = 1' | 24 | 72' × 48' | 0,0 / 72', 48' |
| | 3/4 in. = 1' | 16 | 48' × 32' | 0,0 / 48', 32' |
| | 1 in. = 1' | 12 | 36' × 24' | 0,0 / 36', 24' |
| | 1.5 in. = 1' | 8 | 24' × 16' | 0,0 / 24', 16' |
| | 1 in. = 20' | 240 | 720' × 480' | 0,0 / 720', 480' |
| 42 × 30 | 1/16 in. = 1' | 192 | 642' × 480' | 0,0 / 642', 480' |
| | 1/8 in. = 1' | 96 | 336' × 240' | 0,0 / 336', 240' |
| | 3/16 in. = 1' | 64 | 224' × 160' | 0,0 / 224', 160' |
| | 1/4 in. = 1' | 48 | 168' × 120' | 0,0 / 168', 120' |
| | 3/8 in. = 1' | 32 | 112' × 80' | 0,0 / 112', 80' |
| | 1/2 in. = 1' | 24 | 84' × 60' | 0,0 / 84', 60' |
| | 3/4 in. = 1' | 16 | 56' × 40' | 0,0 / 56', 40' |
| | 1 in. = 1' | 12 | 42' × 30' | 0,0 / 42', 30' |
| | 1.5 in. = 1' | 8 | 28' × 20' | 0,0 / 28', 20' |
| | 1 in. = 20' | 240 | 840' × 600' | 0,0 / 840', 600' |

# SUPPLEMENT 3

# Text Heights in Model Space

## SETTING TEXT HEIGHTS IN MODEL SPACE

The following table shows **Plotted Text Heights** for typical Architectural Scales. Text entered in Model Space will "reduce" to the Scale of the final Plot—just like other elements on your "full size" Model Space drawing. This means that you must compensate for the reduction in size, in accordance with the scale of the final Plot and the desired text height on that final Plot. While modifications can be made during the drawing process, you will need to make a preliminary decision about the Scale of your final Plot before using this chart. Offices, schools, and instructors have different standards for text heights—unless guided otherwise, you might find the following heights serviceable:

Plotted/Final Heights:    General Notes   $3/32$" Room Names $1/8$"    Titles   $3/16$"

| PLOTTED DRAWING SCALE | SCALE FACTOR | PLOTTED TEXT HEIGHT (in.) | TEXT SIZE on DRAWING (in.) |
|---|---|---|---|
| $1/16$ in. = 1'-0" | 192 | $1/16$ | 12 |
| | | $3/32$ | 18 |
| | | $1/8$ | 24 |
| | | $3/16$ | 36 |
| | | $1/4$ | 48 |
| $1/8$ in. = 1'-0" | 96 | $1/16$ | 6 |
| | | $3/32$ | 9 |
| | | $1/8$ | 12 |
| | | $3/16$ | 18 |
| | | $1/4$ | 24 |
| $3/16$ in. = 1'-0" | 64 | $1/16$ | 4 |
| | | $3/32$ | 6 |
| | | $1/8$ | 8 |
| | | $3/16$ | 12 |
| | | $1/4$ | 16 |
| $1/4$ in. = 1'-0" | 48 | $1/16$ | 3 |
| | | $3/32$ | 4.5 |
| | | $1/8$ | 6 |
| | | $3/16$ | 9 |
| | | $1/4$ | 12 |
| $3/8$ in. = 1'-0" | 32 | $1/16$ | 2 |
| | | $3/32$ | 3 |
| | | $1/8$ | 4 |
| | | $3/16$ | 6 |
| | | $1/4$ | 8 |

| PLOTTED DRAWING SCALE | SCALE FACTOR | PLOTTED TEXT HEIGHT (in.) | TEXT SIZE on DRAWING (in.) |
|---|---|---|---|
| $1/2$ in. = 1'-0" | 24 | $1/16$ | 1.5 |
| | | $3/32$ | 2.25 |
| | | $1/8$ | 3 |
| | | $3/16$ | 4.5 |
| | | $1/4$ | 6 |
| $3/4$ in. = 1'-0" | 16 | $1/16$ | 1 |
| | | $3/32$ | 1.5 |
| | | $1/8$ | 2 |
| | | $3/16$ | 3 |
| | | $1/4$ | 4 |
| 1 in. = 1'-0" | 12 | $1/16$ | .75 |
| | | $3/32$ | 1.125 |
| | | $1/8$ | 1.5 |
| | | $3/16$ | 2.25 |
| | | $1/4$ | 3 |
| $1 1/2$ in. = 1'-0" | 8 | $1/16$ | .5 |
| | | $3/32$ | .75 |
| | | $1/8$ | 1 |
| | | $3/16$ | 1.5 |
| | | $1/4$ | 2 |
| 1 in. = 20'-0" | 240 | $3/32$ | 22.5 |
| | | $3/16$ | 45 |

# Quick Leader

## QUICK LEADER

Please read this *entire* handout before beginning your setup. Mastery of these processes is directly related to PRACTICE—doing things over and over and over again. A second part to mastering these processes lies in your ability to use reference materials to guide you to the desired outcomes—this also takes practice. Your instructor is a good and reliable source for help, but he or she wants you to learn how to use the various guides on your own—for when you are working on your own. Do your best to follow the handouts and guides provided in this class and do your best to use text information. Work with fellow classmates to solve problems. Your instructor stands ready to help and guide if you reach a dead end … but give it the "good old college try" before asking for help.

## DISCUSSION:

**Leader Lines** "lead" from a note to a specific item on a drawing. Formats for leaders will vary from office to office—always make sure that you follow the standards of the office in which you are working. The following guidelines are fairly typical, but individual instructors and/or offices may have you make slight adjustments to the format described in this handout. Be flexible, and be consistent in your applications on a drawing or over a set of drawings.

Typically, leader lines align with the right or left side of a note. It is generally understood as poor procedure to terminate a leader at the top or bottom of a note … though on occasion you will find examples where this is done. Leader lines themselves should be thin—like dimension lines—and, typically, should have a "shoulder" where they meet the note. The lines themselves can be curvilinear or segmental (angled), and typically end in a closed, filled arrow. Again, office procedures may dictate a different arrowhead or termination. Be consistent—don't mix curved and segmental leaders on the same drawing.

Text for the leader's note should be justified left, and typically read from the bottom of the sheet.

THIS NOTE
IS JUSTIFIED LEFT

THIS NOTE
IS JUSTIFIED LEFT

Leader lines may extend from the note to multiple entities:

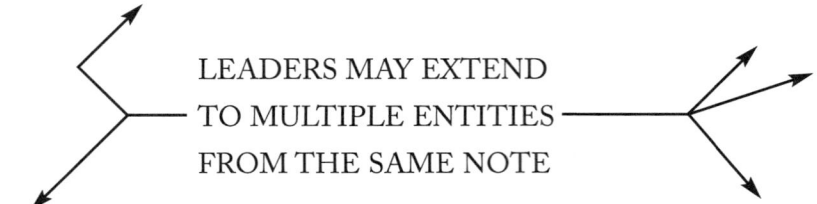

LEADERS MAY EXTEND
TO MULTIPLE ENTITIES
FROM THE SAME NOTE

AutoCAD offers several methods of setting up, accessing, and using a Leader.

This handout will assist in setting up and using the **Quick Leader—qleader.**

## PROCEDURE FOR QUICK LEADER (QLEADER) SETUP

Set a **Text Style** and **Dimension Style** *before* you set and use the Quick Leader.

Arrow sizes, Text formats, and line Color will be determined by those settings.

See the handout for setting Dimension Styles

Type **qleader** and **Enter** (or select icon from Dimension Toolbar)

Reading the Command Prompt line—you have two options:

You can start drawing the leader, if you have completed the setup.

To complete the setup, type **S** for settings and Enter. Do this now.

Type **S** and **Enter** to access the **Leader Settings** dialog box.

Select the **ANNOTATION** tab:

Annotation type = Mtext

Mtext options = Always left justify

Annotation reuse = None

Select the **LEADER LINE and ARROW** tab:

Leader line = Straight

Number of points = 3

Arrowhead = Closed filled

Angle constraints

First segment = Any angle

Second segment = Horizontal

Select the **ATTACHMENT** tab:

Text on left side = Middle of multiline text (verify with instructor)

Text on right side = Middle of multiline text (verify with instructor)

## APPLICATION

From the Dimension Tool Bar select the **leader icon** ... or ...

Type **qleader** and follow prompts.

The first point you select is the location of the ARROW HEAD, the second point will select the shoulder and the third point the shoulder's termination (connection point).

Since you selected three points in your setup, you are now automatically "out" of the line portion and prompted for the text. Type the text—all upper case—Enter at the end of a line to begin another line, Enter a second consecutive time to "stop" writing. Keep notes "compact"—don't let them stretch out across the paper. This is a judgment call on your part. Remember that you should set a text style and dimension style *before* you start with your leader work.

**Note:** If you want **just the leader line** or part of the leader line and no notes, draw the leader segment(s) and then hit the Escape key.

**qleader settings** – Colors, Text Style and Height, Arrow Size—are tied to the Dimension Style. You may wish an arrow size larger than the associated dimensioning "tick." If so, then use **Modify Pulldown/Properties to enlarge arrows**—or set up an alternate Dimension Style for Leaders. If you alter one Arrowhead's size, you can then **Match Properties** to other Arrowheads you wish to change.

# SUPPLEMENT 5

# Floor Plan Linetypes and Layers

## FLOOR PLAN LINETYPES AND LAYERS

Layering formats will vary from office to office. Typically offices will have a template or prototype already in place for plans and other drawing types. Always check for standard templates or formats before beginning to work in an office situation. Use the following formats as a starting point for setting Layers in a Floor Plan drawing file. Final Plots can be "adjusted" by turning various Layers "ON" or "OFF." AutoCAD will display Layers in alphabetical order, so you may wish to designate them in a way that will produce "groups" of related entities. Additional Layers can be created as needed. Plotting formats for the Georgian House and Brownstone units are set by Color—not by Layer—so default Lineweights for Layers are typically "ok." Refer to Supplement 6: Quick Guide to Plotting in Paper Space for Color/Lineweight assignments when Plotting. Your instructor may provide you with alternate Plotting/Lineweight strategies based on Layers—verify settings used in your classroom.

**Load Linetypes: Format** Pulldown/**Linetype/Load** before setting Layers:

> Continuous
>
> Hidden, Hidden 2, HiddenX2
>
> Center, Center2, CenterX2
>
> Phantom

**Global Linetype Scale:** to 24 (for ¼" scaled drawings—see **Supplement 2:**)

**Deselect** "Use paper space units for scaling" box

**Create New Layers: Format** Pulldown/**Layers/New**

**Colors are standard: 1–7, gray 8 and 9.** Colors shown in parentheses recommend different Color assignments for entities on the same Layer. Make these changes as you draw. Since we're assigning Lineweights by Color, start "seeing" different Lineweights when you "see" Colors on your screen.

| | | | |
|---|---|---|---|
| Appliances | Cyan | Continuous | (red details) |
| Doors | Yellow | Continuous | (red swing) |
| Cabinets_Shelves | Blue | Continuous | (green, hidden wall cabs) (green, centerline @ rods) |
| Columns | White | Continuous | |
| Dim_flrpln | Red | Continuous | (green text) |
| Electrical | Cyan | Continuous | |
| Elevator | Magenta | Hidden | |
| Equipment | Magenta | Continuous | (details red) |

(continued)

**(Layers, continued)**

| | | | |
|---|---|---|---|
| Fireplace | White | Continuous | (blue firebrick, blue hearth) |
| Hatch_masonry | Red | Continuous | |
| Hatch_ceramic tile | Gray 8 | Continuous | |
| Hatch_solid fill | Gray 9 | Continuous | |
| Plumbing | Cyan | Continuous | (red details) |
| Mechanical | Cyan | Continuous | |
| Schedules | Yellow | Continuous | (text green, details blue) |
| Steps_concrete | Blue | Continuous | (cyan rails) |
| Steps_ wd-mtl stairs | Cyan | Continuous | (cyan rails) |
| Site_hardscape | Yellow | Continuous | |
| Text_gen notes | Green | Continuous | |
| Text_roomnms | Yellow | Continuous | |
| Text_titles | Yellow | Continuous | |
| Titleblock | White | Continuous | (yellow detail, white/green text) |
| Viewport | White | Continuous | **deselect PLOT** |
| Walls_ext | White | Continuous | |
| Walls_furring | Magenta | Continuous | |
| Walls_half | Blue | Continuous | |
| Walls_int | Yellow | Continuous | |
| Walls_porch-balcony | White | Continuous | (adjust details as needed) |
| Walls_site | White | Continuous | |
| Windows | Magenta | Continuous | (blue jamb, blue frame) |

# Quick Guide to Plotting in Paper Space (2004)

| RM. | PLOTTER | PAPER SIZE | SMALL PLOT | LARGE PLOT | LINE WEIGHT SETTINGS PLOTTER CONFIGURATIONS PAPER SIZES |
|---|---|---|---|---|---|
| | | 8½ × 11 "letter" | × | | |
| | | 8½ × 14 "legal" | × | | |
| | | 11 × 17 "tabloid" | × | | |
| | | 24 × 36 | | × | **FOR SMALL & LARGE PLOTS** |
| | | | | | |
| | | 8½ × 11 "letter" | × | | |
| | | 8½ × 14 "legal" | × | | verify room(s) and plotter(s) |
| | | 11 × 17 "tabloid" | × | | with your instructor |
| | | 24 × 36 | | × | |

| LINE WEIGHTS by COLOR | | | |
|---|---|---|---|
| # | COLOR | SMALL PLOT | LARGE PLOT |
| 1 | RED | 0.05 | 0.15 |
| 2 | YELLOW | 0.20 | 0.60 |
| 3 | GREEN | 0.15 | 0.35 |
| 4 | CYAN | 0.15 | 0.35 |
| 5 | BLUE | 0.18 | 0.45 |
| 6 | MAGENTA | 0.15 | 0.35 |
| 7 | BLACK/WHITE | 0.25 | 0.70 |
| 8 | GRAY | .05 screen 30% | .15 screen 40% |
| 9 | GRAY | .05 screen 20% | .15 screen 20% |

| LINE WEIGHTS BY GRADATION | | | |
|---|---|---|---|
| GRADATIONS | COLOR | SMALL | LARGE |
| Thinnest | RED | 0.05 | 0.15 |
| Medium thin | GREEN | 0.15 | 0.35 |
| Medium thin | CYAN | 0.15 | 0.35 |
| Medium thin | MAGENTA | 0.15 | 0.35 |
| Medium | BLUE | 0.18 | 0.45 |
| Medium wide | YELLOW | 0.20 | 0.60 |
| Widest | WHITE | 0.25 | 0.70 |
| Medium screen | Gray 8 | 0.05 | 0.15 |
| Light screen | Gary 9 | 0.05 | 0.15 |

## PAGE SET UP

1. Left click on a **Layout t**ab to move from Model Space to Paper Space
2. Right click on the same Layout Tab and pick **Page Setup** (typically opens by default)
3. Select the **Plot Device** tab at the top of the **Page Setup** dialog box
4. @ **Plotter Configuration**, select a plotter—see chart above—per room number and desired paper size
5. @ **Plot Style** table..@**Name** ... select **monochrome .ctb** (if you don't have preset lineweights saved and loaded)
6. Select **Edit** to set line widths according to color
7. Set lineweights by selecting the color and then the corresponding value in the **Lineweight** box (@ right)
   Refer to charts above for "small plot" or "large plot" settings/thin, medium, wide lines
   Use the "**Screen**" box to plot a color as a gray tone rather than black
      The smaller the value the lighter the gray, 20–30% is a good place to start
   If you have colors other than the ones shown above, set their lineweights to appropriate widths
   If you want to save these settings for future use, do a SAVE AS with your initials followed by the word
      "small" or "large" according to your settings, i.e., ABCsmall ABClarge
      Create a new folder named **PLOT** in a "safe" location (verify w/instructor), **SAVE AS** to that folder
      To access these settings: Tools/Options/Printer Support Path/Plot Style Table Search Path.
         Select and Remove C:\ ... Browse for and Select your Plot Folder. Apply/OK
      If you don't wish to save these settings for future use, click Save and Close
8. Select the **Layout Settings** tab at the top of the dialog box

9. at **Paper Size:** select the desired paper size—refer to chart above
10. at **Printable Area:** select "**inches**"
11. at **Drawing Orientation:** select **Landscape** or Portrait (landscape is typical for our work)
12. at **Plot Area:** select **Layout**
13. at **Plot Scale** at **Scale:** select 1:1
14. at **Layout Name** (at top) … Delete the "layout1" and rename the setup: 17×11/36×24/Plan/Elevations/etc.
15. Do a **SAVE**—this saves the Layout with its new name.

## VIEWPORT MANIPULATION

1. Check the button display at the bottom of your screen to verify PAPER space or MODEL space mode
2. To toggle from PAPER space to MODEL space:
   Double left click "inside" a viewport for MODEL, double left click "outside" the viewport for PAPER space
   OR use the toggle button at the bottom of the screen clicking between PAPER and MODEL options
3. To erase a viewport: make sure that you are in PAPER space, pick "directly on" one of the viewport lines—erase
4. To create a new viewport: from the viewport tool bar, select Single Viewport, make a window
   Make a NEW layer called "VPORT" for your viewports—click the PLOT symbol "off"
   Make sure that your viewports are on that layer—you will see them on your screen but they won't plot
5. To move a viewport: make sure that you are in PAPER space, pick "directly on" a viewport line, use the MOVE command
6. To change the size of a viewport: make sure that you are in PAPER space, pick "directly on" a viewport line, use grips to resize
7. To manipulate drawing location within a viewport: put the viewport into MODEL space, use PAN command ("Mr. Hand"), right click to exit
8. To SCALE a viewport: "double click" the viewport into MODEL space, select desired scale from the viewport toolbar
9. To "freely" manipulate scale within a viewport: put the viewport into MODEL space, use the REAL-TIME ZOOM command, right clik to exit

**"LOST IN PAPER SPACE"? Verify toggle at the bottom of the screen says PAPER … not MODEL … then ZOOM ALL.**

## SETTING LINETYPES IN PAPER SPACE

To make hidden lines, centerlines, etc. Show in PAPER Space, go to **FORMAT** Pulldown menu at the top of the screen, select **Linetype**

**"Deselect"** (make it "empty") the **"Use Paper Space units for scaling"** box

You may need to "tab" back and forth from MODEL to PAPER space a few times for effect

## INSERTING TITLE BLOCK

The following assumes that you have created a TITLE BLOCK on a separate drawing

With Drawing Area/Limits of 36" × 24"

Orientated with correct margins on the right-hand side of the sheet

In PAPER space—insert Title Block, using the **Insert** command

Browse to find the Title Block

"Deselect" the Insertion Point/Specify On-screen box—coordinates should read *X*:0; *Y*:0

At the Scale settings: leave both *X* and *Y* settings at "1" if you are plotting on a 36 × 24 sheet

for 17 × 11 sheet, set *X* to the fraction 17/36; set *Y* to the fraction 11/24

**Xplode I** (Inherit) or **Explode** the **Title Block** if you need to edit text. Type ED for edit and select text for editing

**Make sure that the Title Block is NOT on the viewport layer—that's not going to plot, remember?**

## PLOTTING

Right click on the layout tab—select PLOT

Do a full preview—Check to make sure that entire plot is captured by the paper, including Title Block

Make sure that viewports are not plotting (unless you want them to)

right click to Exit

Pick **OK**

# Quick Guide to Residential Planning

## QUICK GUIDE TO RESIDENTIAL PLANNING

The following information is provided as a "quick guide" to laying out common elements in a residential floor plan. In many cases they represent rules of thumb and/or minimum standards. *Remember that "minimum" standards are not necessarily the "best," and that each job and each office may have its own set of requirements.* **In an office situation**, **ALWAYS verify these dimensions with your supervisor, and ALWAYS verify required clearances of fixtures and appliances with the manufacturer's written information.**

Interior Spaces:

The dimensions given are **"clear" dimensions,** that is to say, they represent the clear dimension from the interior *face* of one wall to the interior *face* of an opposing wall.

| | |
|---|---|
| ½ Bath (powder room) | 5' × 5' |
| Typical three fixture bath | 5' × 8' |
| Typical clothes closet | 2' deep (length varies) |
| Walk in closet (2 shelves) | 6' wide (length varies) this is TIGHT! |
| Walk in closet (1 shelf) | 5' wide (length varies) |
| Linen closet | 1' deep (deeper—up to 2' better) length varies |
| Circulation (hallways) | 3' minimum, wider is better |
| Bedrooms | 10' × 10' minimum; 12' × 12' good; master bedroom larger |
| | Remember rooms don't have to be "square" |

Cabinets:

| | |
|---|---|
| Base cabinets and vanities | 2' |
| Wall cabinets | 1' |
| Clothes closet shelves | 1' |
| Linen and pantry shelves | Typically full depth of closet |

Appliances/Fixtures:

| | |
|---|---|
| Refrigerator | 3' Typical; 4' Large (rough-in space) |
| Range/oven | 2'-6" Wide (rough-in space) |
| Dishwasher | 2'-0" Wide |
| Trash compactor | 1'-6" Wide (varies) |
| Cooktop (only) | 2'-6" Wide (can get much wider) |
| Oven or double oven | 2' Wide (27" Wide Cabinet) |
| Washer & dryer | 5' Clear (rough in total) |
| Tub/shower | 2'-8" × 5'-0" (typical) 2'-6" × 5'-0" (minimum) |
| Whirlpool tub | Varies—see catalogs |
| Water closet | 15" minimum clear each side of centerline |
| Kitchen sink *cabinet* | 36" wide typical—see catalogs for actual sink sizes |

**Equipment:**

| | |
|---|---|
| Hot water tank | 2'-0" diameter (varies) |
| Air handler (A/H) | 2'-0" × 2'-0" (varies) |
| Compressor | 3'-0" × 3'-0" (varies) |

**Walls:**

masonry = 8" (furring = 2"), exterior wood frame = $6^1/_2$",
interior partitions = $4^1/_2$" (typ.), plumbing = $6^1/_2$" or $8^1/_2$",
1-hour rated wood frame walls = $4^3/_4$",
$1/_2$" gypsum board typical—"green board" in baths
$5/_8$" F.R. type 'X' gyp. bd. required at garage walls/ceiling

**Concrete Masonry Units (CMU):**

When possible, hold to an 8" modular dimension for wall lengths, openings, and distance between openings. The 8" modular rule: odd feet end in 4" (3'-4"; 27'-4", 53'-4")

even feet end in 0" or 8" (4'-0"; 4'-8"; 56'-0"; 124'-8")

**Doors:**

*Standard Doors* are typically available in 2" modules or increments.

Allow a *minimum* of 2" *each side* for the jamb or frame and casing—3" to 4" is better. Doors are typically 6'-8" high. Remember that these are "suggested" sizes. Select sizes for pairs (double) of doors as though one will be "pinned" in place—that is for a pair of front doors, you must select a pair of 36" wide doors … two 24" doors wouldn't cut it.

Standard door widths from "tight" to "extravagant"
(Recommended or "typical" sizes are highlighted):

| | |
|---|---|
| Front | **36"** |
| Utility room | 32" – **34"** – **36"** |
| Bedrooms | 30" – 32" – **34"** (typical) – **36"** (master) |
| Bathrooms | 28" – **30"** |
| Walk-in closets | 26" – 28" – **30"** – 32" |
| Linen closets | 24" |
| *Sliding glass doors.* | **6'** or 8' wide |

*Closet Doors* are typically available in 12" modules or increments—that's for the "entire opening"—and offer the style options of "Bi Pass" (sliding) or "Bi Fold" (fold into a stack). You do not typically need to plan for a jamb or frame for these doors—just create a TOTAL opening in the wall on an increment of 12"—that is 2', 3', 4', 5', or 6'. Larger than 6', go to "two" sets of doors. Other sizes are available—if a pair of Bi Folds are made for a 5' opening, then of course one from that pair would work in a 2'-6" space. The even foot thing is, however, a good basic planning strategy.

**Windows:**

In **Concrete Masonry Unit** (CMU) construction, hold window (and door) openings to block module. That way the masons won't need to cut blocks to create the window opening. This means window widths should fall at 1'-4", 2'-0", 2'-8", 3'-4", 4'-0", 4'-8", 5'-4", 6'-0". Remember that "odd feet" get 4" and "even feet" get 0" or 8". (A 3'-0" door gets a 2" frame on each side, filling a modular 3'-4" opening perfectly.)

In **Wood Frame** construction, windows are "all over the place" with their dimensions—they get really whacky—so for our immediate purposes, hold window widths to increments of 6". If you need absolute accuracy (like, if you're working in an office), then check the manufacturer's catalogs.

Plan for continuity in your fenestration (window layouts). Look for opportunities to use windows of the same size (when appropriate) rather than using a hodgepodge of sizes. Typically, in elevation, head heights align with door head heights (6'-8"), but this is not a hard and fast rule.

Building codes typically require that each *habitable room* have natural light equal to one-tenth the room's floor area, and that each sleeping room have two exits—one of which may be an operable window of specific width and height. In office situations, *always* verify code requirements.

# Student Self-Assessment

| Supplement 8: STUDENT SELF-ASSESSMENT | SEMESTER START | | | SEMESTER MIDPOINT | | | SEMESTER END | | | NAME: _____  SEMESTER: _____  SECTION: |
|---|---|---|---|---|---|---|---|---|---|---|
| **TASKS/COMPETENCIES** | Not in hand | Attained | Mastered | Not in hand | Attained | Mastered | Not in hand | Attained | Mastered | **INSTRUCTOR:** |
| TURN ON COMPUTER | | | | | | | | | | Use the adjacent matrix as a self-assessment tool, gauging your progress through the course of the semester. |
| OPEN ACAD | | | | | | | | | | |
| USE WIZARD/CREATE NEW DWG. | | | | | | | | | | |
| OPEN EXISTING FILES | | | | | | | | | | |
| SAVE/SAVE AS | | | | | | | | | | "X" the box that in your opinion best describes your skill level for each of the described tasks or operations. |
| SHUT DOWN COMPUTER | | | | | | | | | | |
| SET UNITS | | | | | | | | | | |
| SET LIMITS/PAPER SIZE | | | | | | | | | | |
| SET SNAP | | | | | | | | | | **Not in Hand** means that you could not or cannot perform this task or operation, with or without normal supervision. |
| SET GRID | | | | | | | | | | |
| SET POLAR TRACKING ANGLE(S) | | | | | | | | | | |
| SET ORTHO | | | | | | | | | | |
| SET/USE LAYERS | | | | | | | | | | |
| LINETYPES: LOAD/SCALE | | | | | | | | | | |
| SET TEXT STYLE | | | | | | | | | | |
| SET DIMENSION STYLE | | | | | | | | | | |
| LINE | | | | | | | | | | **Attained** means that you could or can perform this task or operation with normal supervision, that is, with a "little reminder" or direction you could do this |
| ERASE | | | | | | | | | | |
| SELECT ITEMS: PICK/WINDOW(S) | | | | | | | | | | |
| OFFSET | | | | | | | | | | |
| FILLET (0) | | | | | | | | | | |
| FILLET RADIUS | | | | | | | | | | |
| UNDO | | | | | | | | | | |
| ZOOM IN/OUT | | | | | | | | | | **Mastered** means that you could or can perform this task or operation with minimal or no supervision |
| PAN | | | | | | | | | | |
| TRIM | | | | | | | | | | |
| OBJECT SNAPS—SELECT | | | | | | | | | | This assessment is **not graded** and in **no way** affects your final grade. It will be used to assess teaching and learning strategies. |
| SET/APPLY RUNNING OSNAPS | | | | | | | | | | |
| GRIPS | | | | | | | | | | |
| BREAK @ POINT | | | | | | | | | | |
| CIRCLE | | | | | | | | | | |
| COPY | | | | | | | | | | |
| MULTIPLE COPY | | | | | | | | | | |
| MIRROR | | | | | | | | | | |
| EXTEND | | | | | | | | | | |
| CHANGE PROPERTIES | | | | | | | | | | |
| MATCH PROPERTIES | | | | | | | | | | |
| ARC | | | | | | | | | | |
| ARRAY | | | | | | | | | | |
| ROTATE | | | | | | | | | | |
| INSERT | | | | | | | | | | |
| EXPLODE/XPLODE | | | | | | | | | | |
| SET/APPLY TEXT | | | | | | | | | | |
| SET/APPLY DIMENSIONS | | | | | | | | | | |
| SET/APPLY QUICK LEADER | | | | | | | | | | |
| HATCH—PATTERNS, SET SCALES | | | | | | | | | | |
| HATCH—SOLID/DISPLAY ORDER | | | | | | | | | | |
| ED (TEXT EDIT) | | | | | | | | | | |
| PAPER SPACE: PAGE SETUP | | | | | | | | | | |
| PAPER SPACE: PLOT to SCALE | | | | | | | | | | |
| DRAW PLANS AND ELEVATIONS | | | | | | | | | | |
| SOLVE PROBLEMS ON OWN | | | | | | | | | | |
| SOLVE PROBLEMS IN GROUPS | | | | | | | | | | |
| APPLY PREVIOUS KNOWLEDGE | | | | | | | | | | |
| PROFESSIONAL WORK HABITS | | | | | | | | | | |